In full
COLOUR

HOW TO PASS

INTERMEDIATE 2

GEOGRAPHY

Dr. Bill Dick

HODDER
GIBSON
PART OF HACHETTE LIVRE UK

Acknowledgements

The Publishers would like to thank the following for permission to reproduce copyright material:

Extracts from Question Papers are reprinted by permission of the Scottish Qualifications Authority.

Extracts from maps on pages 22, 99 and 100 reproduced by permission of Ordnance Survey on behalf of HMSO. © Crown copyright 2006. All rights reserved. Ordnance Survey Licence number 100047450.

Every effort has been made to trace all copyright holders, but if any have been inadvertently overlooked the Publishers will be pleased to make the necessary arrangements at the first opportunity.

For Christine, Duncan, Louisa and Alistair

Although every effort has been made to ensure that website addresses are correct at time of going to press, Hodder Gibson cannot be held responsible for the content of any website mentioned in this book. It is sometimes possible to find a relocated web page by typing in the address of the home page for a website in the URL window of your browser.

Hachette's policy is to use papers that are natural, renewable and recyclable products and made from wood grown in sustainable forests. The logging and manufacturing processes are expected to conform to the environmental regulations of the country of origin.

Orders: please contact Bookpoint Ltd, 130 Milton Park, Abingdon, Oxon OX14 4SB. Telephone: (44) 01235 827720. Fax: (44) 01235 400454. Lines are open 9.00–5.00, Monday to Saturday, with a 24-hour message answering service. Visit our website at www.hoddereducation.co.uk. Hodder Gibson can be contacted direct on: Tel: 0141 848 1609; Fax: 0141 889 6315; email: hoddergibson@hodder.co.uk

© **Bill Dick, 2006, 2008**
First published in 2006 by
Hodder Gibson, an imprint of Hodder Education,
part of Hachette Livre UK,
2a Christie Street
Paisley PA1 1NB

This colour edition first published 2008

Impression number   5   4   3   2   1

Year   2012   2011   2010   2009   2008

Cover photo from BL Images Ltd/Alamy
Illustrations by Tony Wilkins Design and Phoenix Photosetting
Cartoons © Moira Munro 2008
Typeset in 10.5 on 14pt Frutiger Light by Phoenix Photosetting, Chatham, Kent
Printed in Italy

A catalogue record for this title is available from the British Library

ISBN-13: 978-0340-974-124

CONTENTS

INTRODUCTION

Introduction and Examination Advice

During 2003 the Intermediate Geography course and examinations were reviewed by the Scottish Qualifications Authority. The purpose of the review was to bring the syllabus more into line with the revised Higher Geography syllabus. This process allowed candidates presented at Intermediate 2 level to have a much better preparation for and the opportunity to move up to Higher Geography at a later stage.

Consequently there were significant changes to the syllabus content from the previous syllabus and this is reflected in the examination papers at levels 1 and 2. The revised syllabus consists of three sections namely, the Physical Environment, the Human Environment and Environmental Interactions. These are similar to the three sections in the Higher course. The examination was changed to two papers.

Paper 1 consists of two compulsory questions, the first on **Physical Environments**, the second on **Human Environments**.

At Intermediate 1, the two questions have a total of 20 marks each, to give a total of 40 marks.

At Intermediate 2, the two questions have a total of 25 marks each, to give a total of 50 marks.

Paper 2 examines **Environmental Issues**.

At both levels, candidates are required to select two questions from a total of 5 different topics.

At Intermediate 1, these questions have a total of 10 marks each, to give a total of 20 marks. At Intermediate 2, these questions have a total of 15 marks each, to give a total of 30 marks.

This means that the total number of marks for Intermediate 1 papers is 60 marks and for Intermediate 2 the total is 80 marks.

The time allotted to Intermediate 1 and Intermediate 2 is 1 hour 30 minutes and 2 hours respectively.

This revision guide provides you with a handbook which, when used in conjunction with your class notes and textbooks, will help you to maximise the number of marks you can obtain in the national examinations in Intermediate Levels 1 and 2 Geography.

All of the text is written from an examination point of view. Consequently, this guide is not intended to replace your course books but rather to supplement them for revision purposes. There are certain key elements which examiners look for in responses to questions set. It follows that there are certain basic areas of the course material which candidates must know and which should be included in answers.

Much of the content within the text is laid out in bullet points. You do not have to remember every element of every bullet point, as long as you do remember an appropriate number for the number of marks allocated to any given question.

Throughout the text there are sample questions and marked answers with comments in order to give you a guide as to the kinds of questions which could be asked in the external examinations and the types of answers which would attract high marks.

Helping you to become aware of what is necessary and what is non-essential in preparing for the exams is the guiding principle of these notes. It is important to realise that the examinations do not attempt to examine all of the topics contained within the syllabus in the one examination. It is always helpful to look at previous papers to see which topics have been recently examined. This might provide a clue to the topics to which you could give priority when studying, or for the current exam.

Examination advice

Revision and preparation

◆ When revising you can use this guide together with past examination questions and revise topic by topic such as Glaciated Upland for the Physical Environment paper, or World Population Change and growth rates over time for the Human Environment paper.

◆ Use the question and answer sections to revise. Try the questions and check your answers against the marking comments which are included at the end of the sample questions. When you mark your answers, award a mark for every correct statement or appropriate example.

◆ Time yourself as you try the questions so you can learn the pace you need to work at to get through all the questions in the exam.

◆ There is often a certain amount of predictability in the topics being asked each year. Use past papers to review topics previously asked and prioritise your study topics.

◆ Check your knowledge of the topics from time to time using past paper questions.

◆ You may wish to use the mind mapping technique to help you revise individual topics.

◆ Organise your notes into sections which relate to the exam topics. Work out a schedule for studying with a programme and make sure you know which sections of the syllabus you intend to study.

◆ Organise your notes into checklists and revision cards.

◆ Practice drawing diagrams which may be included in your answers, for example corries or pyramidal peaks.

◆ Try to avoid leaving your studies until the night before the examination, and don't stay up late to study.

During the examination

◆ Make sure you know the examination timetable, noting the dates, times and locations of your examinations.

◆ Give yourself plenty of time by arriving early for the examination, well equipped with pens, pencils, rubbers etc.

◆ Read the exam instructions carefully and make sure you know which questions you need to answer.

- Read each question carefully so that you can avoid needless errors such as answering the wrong sections, or perhaps omitting to refer to a named area or case study. If asked to describe and explain make sure that you do actually have descriptions and explanations in your answer.

- Be guided by the number of marks for a question as to the length of your answer.

- If you are given data in the form of maps, diagrams and tables in the question, make sure you use this information in your answer to support any points of view you give. If describing climates, give climate figures.

- If you are asked for a named country or city, make sure you include details of any case study you have covered. Avoid vague answers when asked for detail. Avoid vague terms such as 'dry soils' or fertile soils' if you can give more detailed information such as 'deep and well drained soils' or 'rich in nutrients'.

- Watch your time and do not spend too much time on any particular answer thus leaving yourself short of time to finish the paper. Try to time yourself during the examination for each question. Make sure that you leave yourself sufficient time to answer all of the questions.

- If you have any time left at the end of the exam, use this time productively to go back over your answers to see if you can add anything to what you have already written. This is especially helpful in questions based on Ordnance Survey maps.

- One technique which you might find helpful, especially when answering long questions worth 6 or more marks is to 'brainstorm' yourself for possible points for your answer. You can write these down in a list at the start of your answer. As you go through your answer you can double check with your list to ensure that you have put as much into your answer that you can. It avoids coming out of the exam and being annoyed that you forgot to mention an important point.

Common errors

Markers of the external examination often remark on errors which occur frequently in candidate's answers.

- **Lack of sufficient detail:** This often occurs in Intermediate 2 answers, especially in 5 or 6 mark questions. Many candidates fail to provide sufficient detail in answers often by omitting reference to specific examples, elaborating or developing points made in their answer. Remember a good guide to the amount of detail required is the number of marks given for the question. If, for example, the total marks offered is 6, then you should make at least six valid points. At Intermediate 1 the number of lines provided for the answer indicates the appropriate length of answer required.

- **Listing:** If you give a simple list of points rather than fuller statements in your answer you will automatically lose marks. For example, in a 4 mark question you will obtain only 1 mark for a list.

- **Bullet points:** The same rule applies to a simple list of bullet points. However if you give bullet points with some detailed explanation you could achieve full marks.

- **Irrelevant answers:** You must read the question instructions carefully to avoid giving answers which are irrelevant to the question. For example, if asked to explain and you simply describe you will lose marks. If asked for a named example and you do not provide one, you will lose marks.

◆ **Repetition:** You should be careful not to repeat points already made in your answer. These will not gain any further marks. You may feel that you have written a long answer, but it may contain the same basic information repeated again and again. Unfortunately these repeated statements will be ignored by the marker.

Chapter 1

PHYSICAL ENVIRONMENTS

Landscape Locations and Types

Key Idea 1

You should be able to locate four different landscape types, namely, glaciated uplands, upland limestone, coastlines of erosion and deposition and rivers and their valleys within the regional context of the British Isles. The locations which will be used in assessments are shown on the maps in Figures 1.1, 1.2 and 1.3.

Key Idea 2

You are expected to be able to recognise, describe and explain the formation of the main features of these landscapes.

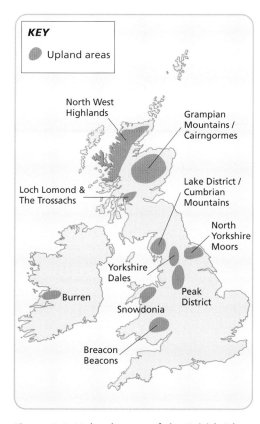

Figure 1.1 Upland areas of the British Isles

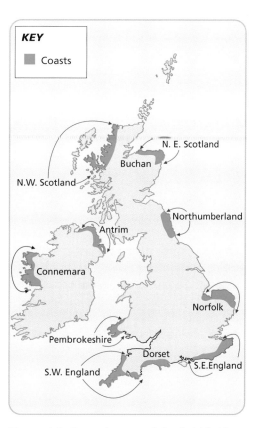

Figure 1.2 Coastal areas of the British Isles

HOW TO PASS INTERMEDIATE 2 GEOGRAPHY

Figure 1.3 Main rivers of the British Isles

 Landscape type 1 Glaciated upland

Key Point 1

You should know the main processes which helped to form glaciated upland.

At various times during the last 2.5 million years the British Isles has been covered with large sheets of ice. These periods are called glaciations and there may have been up to 20 different glaciations during the period known as the Ice Age. The last Ice Age ended about 10 000 years ago. During the Ice Age, ice, in the form of glaciers, advanced southwards as the climate became colder and retreated as the temperature gradually became warmer. These advances and retreats are termed glacial and interglacial periods, respectively.

Formation processes

◆ Glaciers consist of rivers of snow and ice which flow over *bedrock*. The weight of the ice causes the glacier to slide on top of the melting ice. This is called basal sliding.

◆ Melting ice near the base of the glacier causes a process within the glacier known as internal deformation.

◆ The rate of flow of the glacier depends on the type of rock over which it flows, the amount of ice in the glacier and the slope of the land.

◆ As glaciers move they erode and deposit material at their margins (the areas at the front and sides of glaciers).

◆ Snow and ice are lost from the glacier through melting and erosion. This loss of snow and ice is called *ablation*.

◆ Glacial erosion occurs through two processes, *abrasion* and *plucking*.

◆ Abrasion occurs when small rock debris in the ice slides over the bedrock and smoothes it away. Abrasion produces smooth surfaces.

◆ Plucking occurs when changes in the flow conditions cause the thin film of meltwater between the glacier and the bedrock to freeze. As the glacier moves, fragments of rock are ripped or plucked from the ground. Plucking produces jagged features.

◆ Meltwater flowing from the glaciers further eroded and deposited material in a process called fluvioglacial processes.

Key Point 2

Erosion by glaciers produces several post-glacial landforms. You should be able to identify these landforms, describe their main characteristics and explain how they were formed. These features include corries, tarns (or corrie lochs), pyramidal peaks, arêtes, U-shaped valleys, misfit streams, truncated spurs, hanging valleys and ribbon lakes. These features are shown in Figure 1.4.

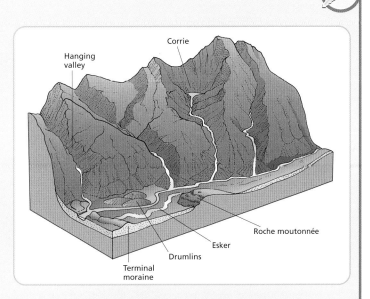

Figure 1.4 Upland glaciated landscape

Formation of post-glacial features

Corries

◆ Corries are steep-sided hollows in the sides of mountains where snow has accumulated and gradually compacted into ice.

◆ The rotational movement of ice in the hollow caused considerable erosion both on the floor and on the sides of the depression.

◆ The erosion on the floor was caused by abrasion and the floor becomes concave in shape and the edge takes on a ridge-shaped appearance.

◆ As the corrie filled up with ice eventually it could not contain any more ice and some of it moved down the slope to a lower level. This was the beginning of a glacier.

◆ At the sides of the corrie, plucking of rocks took place as the ice moved forward and the back wall of the depression became very steep.

Tarns (or corrie lochs)

◆ Occasionally as the ice melted, melt water filled the corrie forming a lake called corrie lochs (Scotland) and tarns (England).

Arêtes

◆ Corries often developed on adjacent sides of a mountain and when they were fully formed they were separated by a knife-shaped ridge termed an arête.

Pyramidal peaks

◆ If corries develop on all sides of a mountain, the arêtes will form a jagged peak at the top. This feature is called a pyramidal peak. These are further sharpened by frost action.

U-shaped valleys

◆ As a glacier moved downhill through a valley, the shape of the valley was transformed. A material called boulder clay is deposited on the floor of the valley.

◆ As the ice melts and retreats the valley is left with very steep sides and a wide flat floor, in the shape of a letter U.

Misfit streams

◆ Meltwater from a glacier may form a river or stream flowing through U-shaped valley.

◆ This replaced the original stream or river and is termed a 'misfit' stream.

Truncated spurs

◆ A spur is the bottom part of a ridge which juts out into the main valley. These spurs were removed as the glacier cut through the valley.

◆ The feature remaining once the ice melts is called a truncated spur.

Hanging valleys

◆ The sides of the U-shaped valley are usually high and steep, and have tributary valleys feeding into them. During the Ice Age, these tributary valleys often held smaller glaciers.

◆ The glacier in the main valley cut off the bottom slope of the tributary valley leaving it high above the main valley.

◆ Tributaries of the main valley will therefore plunge from the slopes of the main valley into the bottom of the valley. These smaller valleys are called hanging valleys.

Ribbon lakes

◆ The material which was pushed in front of the glacier and left as the glacier melted is called *terminal moraine*.

◆ This material may be large enough to form a dam at the end of the U-shaped valley.

◆ The stream cannot proceed further and gradually backs up. The valley is then flooded and the shape of the resulting lake or rock resembles a piece of ribbon stretching back through the length of the valley.

◆ These bodies of water are called ribbon lakes (in England) or ribbon lochs (in Scotland). Examples include Ullswater, Windermere and Coniston Water in the Lake District in North West England and Loch Lomond in West Central Scotland.

Questions and Answers

Question 1.1

Explain how one of the following features of an upland glaciated area was formed. You may use diagrams in your answer. Choose from: *arête*; *hanging valley*; *U-shaped valley*; *corrie*.   *(3 marks)*

Answers

Arête

An arête is formed when two glaciers are about to meet (✓). No abrasion is used to form this, only plucking and frost shattering (✓). The arête is usually very steep with scree at the bottom.

Comments and marks obtained

Without the diagrams showing the processes and stages of development worth a mark in itself, this answer would only have gained **9** marks as indicated. The last sentence describes rather than explains and is therefore not worth any further marks. With the diagrams the answer is worth **3 marks out of 3.**

Corrie

Ice and snow gather in a hollow (✓) and as more gathers the bottom layers are compressed by the weight (✓) and begin to erode the hollow (✓). When the ice and snow finally melt a steep back wall is left and the original hollow is much deeper (✓). As there is nowhere for the water to drain away to, a loch called a tarn is formed in the corrie (✓).

Comments and marks obtained

This answer is quite basic but it contains enough detail especially about the ice and snow being compressed and eroding a steep back wall and melting ice left to form a corrie loch or tarn. A better answer might have referred to 'plucking and abrasion'

Questions and Answers continued ➢

Questions and Answers continued

and the rotational movement of the ice in the hollow helping to create the back wall and a 'lip' at the front edge of the corrie. However, the answer manages to get **4 marks out of 4.**

U-shaped valley

A U-shaped valley was formed by a glacier moving down an area of land (✓).

Plucking and abrasion (✓) takes place here eroding the land as the glacier moves down the landscape (✓). Eventually the glacier disappears and the landscape is left as shown.

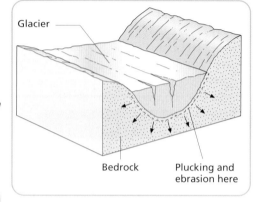

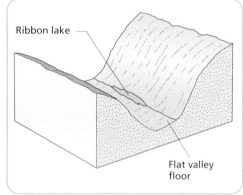

Comments and marks obtained

The opening statement on the movement of the glacier is worth one mark. A further mark is gained for the diagram showing the processes taking place as the glacier moves down the valley. The reference to 'plucking and abrasion' and erosion of the valley gains a further two marks and the final mark is obtained by the diagram showing the final stage of the formation of the U-shaped valley.

The answer obtains a full **4 marks out of 4.**

Landscape type 2 Upland limestone

Key Point 3

You should be able to recognise and describe features of an area of upland limestone. For the purpose of the external examinations these features include limestone pavements, grykes, clints, scree slopes, potholes and swallow holes, caverns, stalactites, stalagmites and intermittent drainage. You should be able to describe and explain the processes which led to the formation of these features.

Figure 1.5 shows a summary diagram of the features associated with areas of upland limestone.

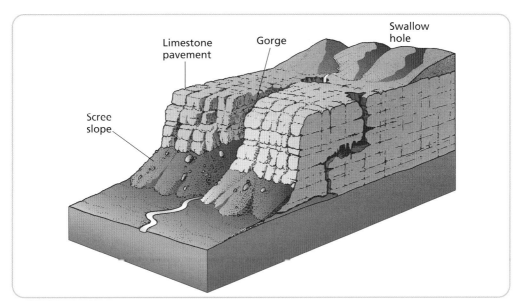

Figure 1.5 Features of an area of upland limestone

Upland limestone

Limestone is a sedimentary rock consisting mainly of calcium carbonate (usually at least 80%). Depending on its age it can form several different land forms. Carboniferous limestone was formed about 250 million years ago and the landscapes have specific features which are immediately recognisable.

Limestone pavements
◆ When glaciers passed over the top of an upland limestone area, the top-soil was removed leaving an area of exposed rock.
◆ The subsequent chemical action of rainwater dissolved the limestone; joints widened and deepened on the surface creating large blocks resembling pavements.

Grykes and clints
◆ The cracks or fissures between the blocks are called grykes and the large blocks separated by the grykes are called clints.

Scree slopes

◆ The edge of the limestone area which is exposed to the elements can form a steep slope known as a scar or scarp slope.

◆ Frost shattering occurs when water seeps into little cracks on the slope surface and freezes during cold winters This causes pieces of rock breaks off and falls down to the bottom of the slope.

◆ This broken material which gradually builds up on the slope is called scree and the slope is called a scree slope.

Potholes and swallow holes

◆ Potholes and swallow holes are formed where persistent widening of a major joint has occurred through weathering by a surface stream which has since disappeared underground.

Underground caverns

◆ As the process of dissolving the limestone continues, underground sections of the rock may collapse creating underground caves.

◆ As the surface water meets the impermeable underground rock, this can lead to the creation of underground lakes and streams.

Stalagmites and stalactites

◆ Stalactites and stalagmites are formed underground in caverns.

◆ Through chemical weathering particles of limestone are dissolved in solution by rainwater.

◆ This water percolates through the rock and drops are deposited on the ceiling and floor of caverns.

◆ Gradually the moisture is evaporated and the deposits of limestone are left.

◆ Stalactites are the deposits of limestone left hanging down from the ceiling of the cave.

◆ Deposits which build up from the ground are called stalagmites.

Intermittent drainage

◆ Intermittent drainage occurs on limestone areas when streams which drain areas of impermeable rock carry on into the limestone area and disappear through the permeable limestone. This interrupts the course of the stream on the surface.

◆ The streams flow along the bedding planes underground until they reach the underlying permeable rock. Gradually as the streams flow along the water table they emerge at the surface at a lower level and this water is called a spring.

◆ Areas of upland limestone in Britain form a rolling plateau-like landscape which has no surface drainage.

◆ Due to the lack of water, vegetation is sparse or non-existent. Exposed hard grey limestone is clearly seen on the surface. This landscape is referred to as Karst scenery.

Gorges

◆ As water passes through the limestone the rock is dissolved. Over time the joints and bedding plains become wider and caverns form.

◆ As the caverns increase in size, the amount of rock above it decreases. Eventually the rock becomes unstable and collapses into the cavern. This forms a deep sided valley or gorge.

◆ The best known gorge of this type in the British Isles is probably Cheddar Gorge in the Mendip Hills of England.

Questions and Answers

Question 1.2

Study Figure Q1.2. Choose one surface feature and one underground feature. Explain how each was formed. You may wish to use diagrams in your answer. *(6 marks)*

Intermediate 2 2003 Q1b

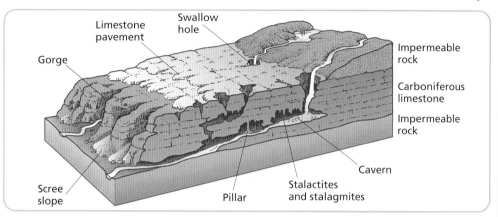

Figure Q1.2 Surface and underground features

Answer

In limestone there are cracks running all the way through it. Since limestone is permeable when rain falls it is absorbed into the rock. But when the rain falls through the atmosphere it picks up fumes from car exhausts and turns it into a weak acid (✓). So when this rain falls on the limestone it dissolves some of it (✓). When this is repeated many times those little cracks grow wider and deeper (✓). The cracks are called grykes and the flat pieces of limestone are called clints (✓). This is how a limestone pavement is formed.

A limestone pillar is formed when the rainwater containing the limestone enters a cavern. Sometimes the rain water evaporates leaving the deposit of carboniferous limestone (✓). When the process is repeated thousands of times a stalactite is formed (✓). Stalagmites are formed in the same way but instead of growing down from the cavern roof they grow up from the cavern floor (✓). Sometimes the stalactites and stalagmites meet in the middle (✓).

Comments and marks obtained

This answer selected two features for explanation, limestone pavements (surface) and limestone pillars (underground). Explanation was provided without the aid of diagrams which is perfectly acceptable.

In the first part of the answer, marks are gained for references to formation of weak acid, acid dissolving limestone, opening of cracks and defining these cracks as clints.

In the explanation of the formation of the pillar, marks are obtained for reference to water evaporating and leaving behind deposits of limestone, repetition of this process forming stalactites and stalagmites and the eventual joining together of these two features. In total the answer has sufficient points to gain 8 marks but the total for the question is 6 and therefore the answer merits **full marks**.

Landscape type 3 Coastlines of erosion and deposition

Key Point 4

You should be able to recognise, describe and explain features of landscapes of coastal erosion and deposition. For examination purposes these features include cliffs, caves, arches, stacks, headlands and bays, beaches, spits, bars and tombolos.

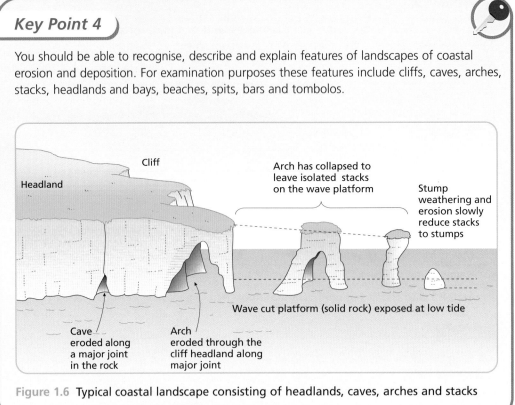

Figure 1.6 Typical coastal landscape consisting of headlands, caves, arches and stacks

Formation of coastal features formed through erosion

Coastal features formed by erosion include cliffs, headlands and bays, caves, arches and stacks. Figure 1.6 shows a typical coastal landscape created by coastal erosion.

Cliffs

◆ Cliffs are formed by wave action undercutting land which meets the sea. This occurs at about high tide level. A notch is cut and as the land recedes the cliff base is deepened by wave erosion. At the same time the cliff face is continually attacked by weathering processes and mass wasting such as slumping occurs causing the cliff face to become less steep.

◆ When high, steep waves break at the bottom of a cliff, the cliff is further undercut forming a feature called a wave cut notch. Continual undercutting causes the cliff to eventually collapse and as this process is repeated the cliff retreats leaving a gently sloping wave cut platform. The slope angle of this is less than 4 degrees.

Headlands and bays

◆ When resistant rocks alternate with less resistant rocks along a coast and are under wave attack, the resistant rocks form headlands whilst the less resistant rock is worn away to form bays.

◆ Headland and bays can also develop in a single rock structure such as limestone, which has lines of weakness such as joints or faults.

◆ Although the headlands gradually become more vulnerable to erosion, they still protect the adjacent bays from the effects of destructive waves.

Caves, arches and stacks

◆ Caves are formed when waves attack cliffs with resistant rock along lines of weakness such as faults and joints.

◆ The waves undercut part of the cliff and can cut right through the cave to form an arch.

◆ Continual erosion causes the arch to widen and eventually the roof of the arch collapses to leave a piece of rock left standing called a stack.

Features formed through coastal deposition

Beaches

◆ Beaches are formed from material which is deposited by the sea. They form a buffer zone between the coast and the sea.

◆ The beach consists of loose material and its shape can be changed by factors such as wave energy, the gradient of the slope, and the seasons of the year.

◆ There are usually two types of beach profile. Beaches are generally either relatively wide and flat or are steep and narrow.

◆ Wave action can be constructive or destructive and beaches can therefore be built up or destroyed by waves.

◆ The type of particle deposited, either sand or shingle, can also affect the shape and width of a beach.

◆ Another process which affects beach formation is the movement of material along the coast by the process known as *longshore drift*. This is the process by which waves carry material up and down a beach. This material is usually deposited in a zig-zag fashion due to the effects of winds on waves.

◆ To prevent the movement of beach material away from beaches by longshore drift, wooden or stone walls called groynes are built along beaches. Although groynes protect some parts of beaches, areas on either side are often depleted of sand.

Spits, bars and tombolos

◆ Spits consist of a long narrow accumulation of sand or shingle with one part still attached to the land, resulting from marine deposition.

◆ The part of the spit not attached to land projects at a narrow angle into the sea or across an estuary. This end is often hooked or curved. Spurn Head on the Humber estuary is a good example of a spit.

◆ Bars are ridges of sediment formed parallel to the coast and can be exposed at low tides.

◆ Tombolos are bars which connect an island to the mainland to form a barrier across a bay. A very well known example of a tombolo in Britain is Chesil Beach near Weymouth.

Questions and Answers

Question 1.3

Using one or more diagrams, explain the formation of a headland. *(4 marks)*

Intermediate 2 2003 2ai

Answer

Headlands are formed of hard rock e.g. granite. These are eroded slowly (✓) while rock in between them is soft e.g. chalk – this is eroded quickly (✓) (forming a bay).

Soft rock in bay is eroded faster. Headland has harder rock so it is eroded faster. So it is left jutting out (✓).

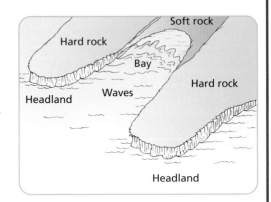

Comments and marks obtained

The first two comments on slow erosion of the granite rock and the faster erosion of the softer chalk gains two marks. The diagram showing alternating rock formation and wave action gains a third mark. The repeating statement on erosion of hard and soft rock gains no further mark, but the final point on the headland being left 'jutting out' gains a final marks. The answer obtains **4 marks out of 4**.

Landscape type 3 Rivers and their valleys

Key Point 5

You should be able to recognise, describe and explain the formation of the main features of a river and its valley. For the external examinations these features are associated with the three stages of development of a river valley namely, the upper, middle and lower stages.

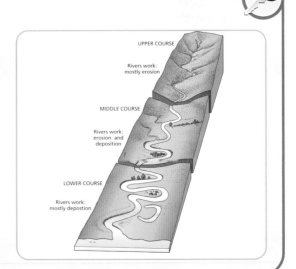

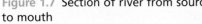
Figure 1.7 Section of river from source to mouth

The basic system of surface drainage is known as a river basin. The system returns water to the oceans and seas as part of the hydrological process.

◆ River valleys can take different shapes often depending on their stage of development.

◆ In the upper stage of development or 'youthful stage', the gradient or slope is usually very steep and the river is fast flowing. The sides of the valley will be steep and at this stage the main work of the river is usually erosion.

◆ The erosion process is greatest during periods of heavy rainfall. The river has more energy to affect the bed load, and material is rolled and bounced along. Downward or vertical erosion occurs and the valley takes on the characteristic V shape.

Key Point 6

You should be able to identify, describe and explain the formation of features in the upper stage of rivers. These features include V-shaped valleys and waterfalls, as shown in Figure 1.8.

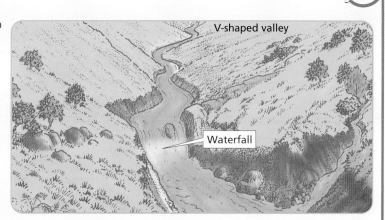

Figure 1.8 V-shaped valleys and waterfalls

V-shaped valleys

◆ In the upper stage of the valley the main process-taking place is erosion.

◆ The river erodes downwards by abrasion and corrosion.

◆ The stream cuts through rock forming a deep channel with steep sides. As the valley deepens the sides become steep and unstable.

◆ Rocks tumble and slide to the bottom of the valley. Loose rock is carried away by the stream.

◆ The valley gradually widens to a V shape.

Waterfalls

Figure 1.9 shows the stages in the formation of a waterfall.

◆ If a river flows over hard rock and then over a band of less resistant rock in a waterfall, the less resistant rock will be worn away much more quickly than the overlying rock.

◆ As the water falls into a plunge pool there is great turbulence causing corrosion and abrasion from rocks swirling about, eroding the pool and making it deeper.

◆ Eventually so much of the underlying rock may be eroded that there will be nothing left to support the rock above. The overlying rock then collapses.

◆ As the process is repeated the waterfall will retreat and this may eventually lead to the formation of a steep sided gorge. A gorge is a deep valley with very steep sides and a narrow valley floor.

◆ Waterfalls may also be formed through glacial erosion, earth movements and a fall in sea level.

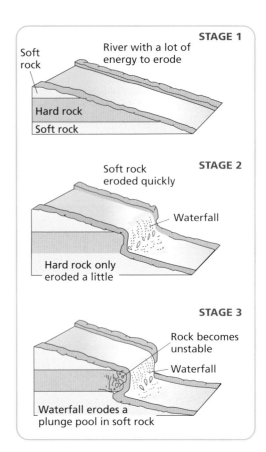

Figure 1.9 Stages in the formation of a waterfall

Key Point 3

You should be able to identify, describe and explain features in the middle and lower stages of rivers. These features include meanders, ox-bow lakes, floodplains and levees.

Meanders

In the middle and later stages the valley sides are less steep and the gradient is more gentle. The width of the river increases and bends or meanders begin to form.

◆ In the middle stage of the valley, the valley sides are less steep although they may still be hilly.

◆ The gradient becomes more gentle and the width of the valley increases with an increase in flat land along the sides of the river.

◆ River bends or meanders begin to appear as the river finds the course of least resistance. Lateral erosion erodes the banks of the channel allowing the meanders to develop.

◆ The speed of flow of the river varies across the meander. The rate of flow is much faster on the outside bend of the meander and at this point the water is eroding the bank.

- On the inside bend the speed of the river is slower and the river deposits material which it is carrying as its load.
- In the lower or final stage of the valley the river widens and flows more slowly across the land.

Ox-bow lakes

- In the lower stages of a river valley as meanders increase in size they may reach a point where extreme loops develop across the floodplain. These loops resemble a horseshoe in shape.

Figure 1.10 Meander and ox-bow lake

- There is a gap between two parts of the horseshoe which becomes very narrow.
- At this narrowing in the meander, the river can cut though the gap cutting off the meander as the river assumes a straighter, more efficient course.
- This separated part of the meander is left as a lake known as an ox-bow lake.

Floodplains

- Floodplains are the areas in the middle and lower stages on either side of a river. The are formed through the deposition of material which the river is carrying.
- When the river overflows on to this area of the valley during a flood, material is deposited on the land as the water drains away.
- This material is called *alluvium* and it gradually builds up to form floodplains.

Levees

- Levees are high raised banks on the side of a river in its lower stage.
- They are formed by the river depositing material during times of flood when it overflows its banks.

Ordnance Survey Maps

Key Idea 3

You should be able to recognise the features of the four landscapes discussed above from photographs, sketches and OS maps.

Upland glaciated features on an Ordnance Survey map

On a landscape diagram or an Ordnance Survey map, either on the scale 1:50 000 or 1:250 000, you should be able to recognize from the contour patterns the features which are formed by erosion. You can be asked to identify and then describe how these features were formed. The level of detail which you need to provide in your answer depends on whether you are attempting either the Intermediate 1 or 2 examination.

If the Ordnance Survey map is based on one of the upland glaciated areas of Britain such as Snowdonia, the Cairngorms or the Lake District, the question often offers a choice of features and may ask you to first of all identify them and then select one of them for detailed description of its formation.

This kind of question could also be based on a block diagram such as that shown in Figure 1.11.

A guide to the amount of detail required should be taken from the number of marks allotted to the question. This usually varies between 3 and 4 marks at levels 1 and 2, with 1 mark given for each correct statement.

You may also be asked to provide appropriate diagrams in your answer and it is quite possible to gain full marks for a well annotated (labelled) series of diagrams. Processes which you should refer to in your answer include abrasion, plucking, rotational movement of ice, transportation and deposition.

Your standard class notes and textbooks will already have provided you with detailed descriptions and explanations of the features mentioned above.

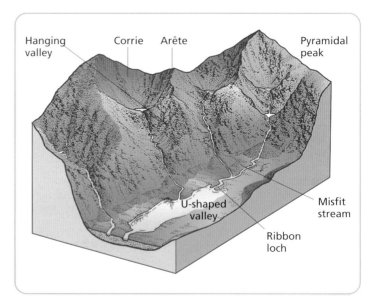

Figure 1.11 Upland glaciated landscape

Landscape features on an Ordnance Survey map

Study the sketches in Figure 1.12. Compare the sketches of the landscape features and the matching contour patterns on an Ordnance Survey map. Try to understand what the contour patterns mean and learn to recognise the various features shown. Study these diagrams and contour patterns and when revising, try to identify similar patterns shown on a variety of map extracts.

In the examination at both levels 1 and 2, you may be asked to either identify the features from a given list or at Intermediate 1 to annotate (or label) the sketch provided.

As well as identifying the features from the maps, you will probably also be required to name examples of these features and to provide appropriate four and six figure grid references.

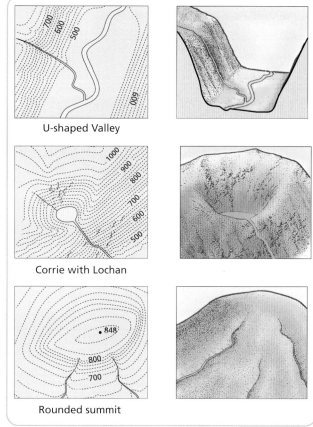

Figure 1.12 a) U-shaped valley; b) Corrie with lochan; c) Rounded summit

Recognising and describing river and river valley features on an OS map

Many of the questions set at both Intermediate 1 and 2 on rivers are based on Ordnance Survey maps. The question may ask you to locate a section of a river by giving two 6 figure grid references.

When describing the features of a river and its valley on an Ordnance Survey map you should refer to:

◆ The stage of development of the valley. Do the contour patterns of the valley suggest a young, mature or old valley? Are the valley sides steep or gentle?

◆ The direction of flow. This should be noted as indicated by both contour lines and spot heights at various points along the course. The gradient should be described as steep or gentle. Use spot heights and contours as a basis for deciding on steepness. Use the gradient of the valley to comment on the width of the river.

◆ Note distinct features within the valley such as meanders, tributaries, ox-bow lakes, braiding and floodplain width.

◆ Refer only to physical features and give appropriate 4 or 6 figure grid references. Do not refer to manmade features such as bridges, roads or settlements.

Questions and Answers

Question 1.4

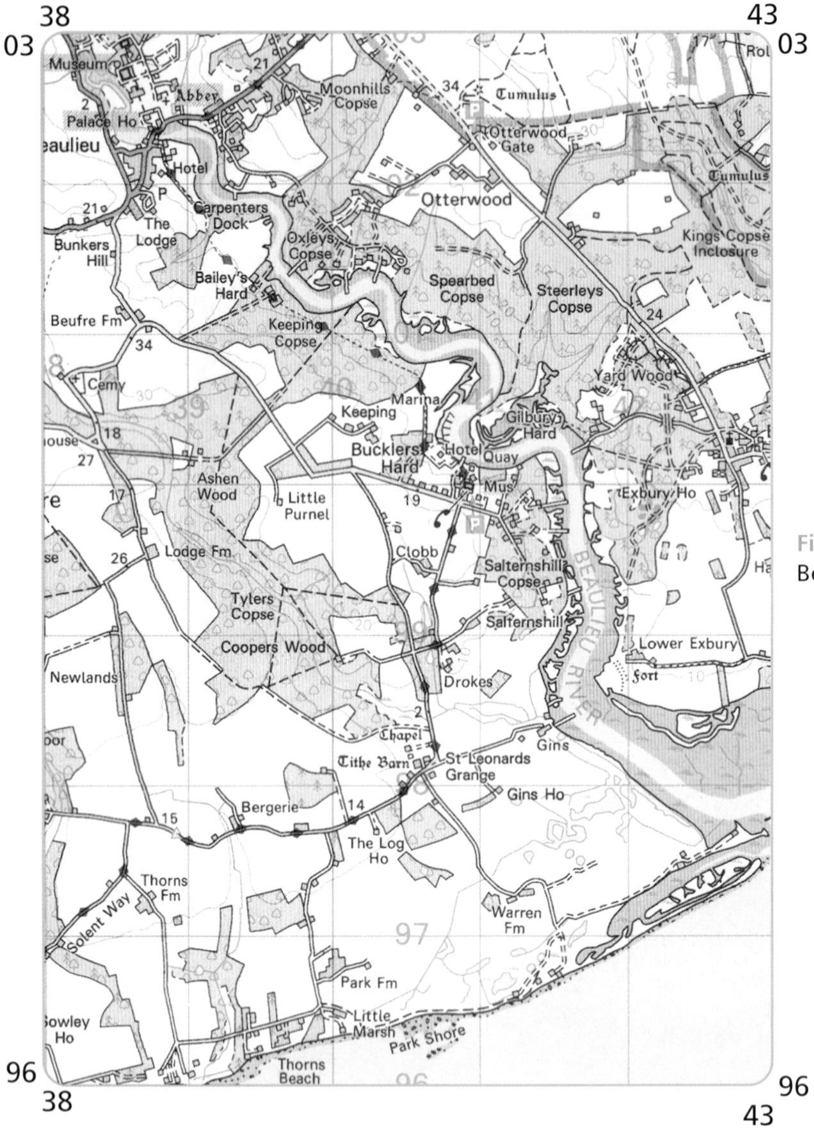

Figure Q1.4 The Beaulieu River

Referring to the OS map, give a detailed description of the physical features of the Beaulieu River and its valley from grid reference 383030 to the river mouth in square 4298. *(4 marks)*

Intermediate 2 2005 1biii

Questions and *Answers continued* ➤

Questions *and* Answers *continued*

Answer

The Beaulieu river is in its lower course and flows through quite a flat area(✓). It flows in a South Westerly direction (✓) and is tidal up to grid square (388023)(✓). It has quite a lot of meanders (403014)(✓). In the tidal part of the river there are a lot of mud deposits. It has six tributaries one of which has a confluence (✓) at (405013).

Comments and marks obtained

Identifying the stage of the river and indicating the steepness of the valley in the first sentence is worth a mark. A second mark is obtained for the comment on the direction of flow and a third for noting that the river is tidal up to grid square 388023. A further mark is gained for the statement on meanders. If the answer required a further mark it would get this for the final statement on the tributaries and confluence point. The answer therefore obtained full marks, **4 out of 4**.

Question 1.5

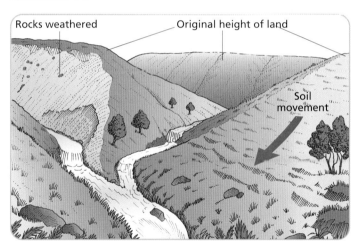

Figure Q1.5 Field sketch of the upper course of a valley

Look at Figure Q1.5 above. Explain how the V-shaped valley was formed.   *(4 marks)*

Questions *and* Answers *continued* ➤

Questions and Answers continued

Answer

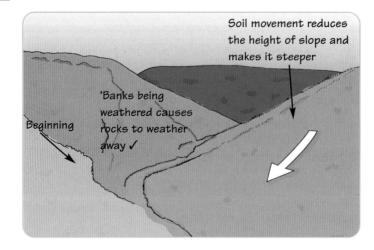

Comments and marks obtained.

Although the candidate uses a diagram they do not give much information. Only the middle comment is sufficiently accurate to obtain a mark for 'weathered away'. Unfortunately the answer does not explain the processes involved and does not refer to erosion as the main factor in the formation of the valley, as opposed to transportation and deposition.

This answer obtains only **1 mark out of a possible 4**.

Case Studies – Land Use

Key Idea 4

Through a case study of one of each of the selected areas of upland glaciation, upland limestone and coastal erosion and deposition you should be able to show knowledge and understanding of land uses appropriate to the areas studied.

Key Point 8

You should be familiar with the land uses in relation to different landscape types.

Farming

◆ There are several different types of farming in Britain and each type has a close relationship with the physical environment. Figure 1.13 indicates the distribution of the main farming types throughout Britain.

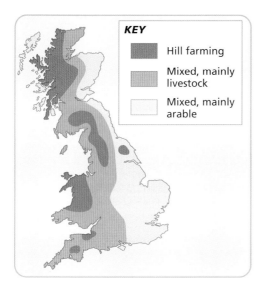

Figure 1.13 Distribution of main farming types throughout Britain

- The type of farming in any area will also be affected by the climate. Crops may flourish if there is plenty of rainfall, or die if there is too much or not enough.

- Temperature changes throughout the year determine growing seasons.

- Farmers may opt for livestock farming as opposed to crop farming due to the influence of climate.

- The flatness of land, steepness of slope and the relative fertility of the underlying soil also encourage certain types of farms and limits others.

Forestry

- Large plantations of commercial forestry are found throughout many parts of Britain, especially in upland areas with steep slopes and areas where soil quality may be relatively poor.

- Forestry is generally restricted to such low-quality land because in areas where the physical conditions are better, the land will be too valuable to be used for forestry, which is a relatively low-value type of farming. This land may be more suited to settlement, industry or farming.

- Over 75 000 tons of timber is produced from commercial forestry each year.

- Since the 1920s, the Forestry Commission has planted large areas of Britain with coniferous trees such as pine, fir, spruce and larch.

- Forests cover 2.8 million hectares of the UK, or about 12% of the total land area. Only 2% of the forest land is semi-natural, and 98% is made up of plantations. These plantations provide wood for the forest products industry in the UK.

Industry

- Industrial land use can vary from primary industries such as mining and quarrying, large heavy industrial complexes such as iron and steelworks and light industrial estates. It may also include power stations such coal- and gas-fired plants and hydro-electric power schemes in upland areas.

- The sites chosen for each of these industrial complexes are greatly influenced by physical factors such as the height and shape of the land, whether there is a water supply nearby and the nature of the underlying rock strata.

- Human factors which influence industrial location include power, labour, transport, capital and markets.

In the examination, you may be shown certain industrial environments in a diagram or on an OS map and be asked to determine the main physical and human factors which influence industrial land use.

Military

◆ Due to a combination of location, terrain and seasonal climatic conditions, some areas of the country are used for military training purposes.

◆ Some areas are very useful for training soldiers and testing equipment. Most of these areas are remote and far from areas used by people. These tend to be in areas of difficult terrain such as mountains and open moorland.

◆ Other areas may be cordoned off for the use of weapons training since proximity to public areas would present a danger to the general public. Such areas include parts of Salisbury Plain and Dartmoor and various coastal parts of Britain.

◆ Occasionally this land use comes into conflict with other land uses, particularly farming and tourism.

Recreation and leisure

◆ Recreational pursuits such as hill walking, climbing, sightseeing, observing nature and sporting activities such as sailing and golfing are closely linked to the nature of the physical environment.

In many questions based on this topic there may not be a right or wrong answer, but the marks you gain will depend on how well you can relate individual factors to specific land uses from a given resource. You have to be able to use the resources given in questions to draw conclusions on the influence of certain physical factors on specific human activities. Often much of this is based on common sense and your skill in map interpretation.

Water storage and supply

◆ Many of the lochs and lakes in Britain are used for water storage and the supply of water to towns and cities and hydroelectric power stations.

◆ Local climate and underlying geology are very important. Water storage and supply areas are located mainly in areas of high average rainfall such as the north and west of Britain. The rock must be impermeable to allow storage reservoirs to hold water.

Tourism

◆ Tourism is one of the most important service industries in Britain. The industry caters for the needs of hundreds of thousands of people who wish to visit places for the purpose of recreation and leisure.

◆ The industry is a source of income for a wide array of jobs, both directly related to employment such as hotels and other forms of accommodation, transport, entertainment and catering to less directly related employment such as health and education services.

◆ Tourists have many different reasons for visiting other areas including recreation, adventure seeking, education, visiting places of historical and cultural interest and sightseeing.

◆ In Britain there are many types of tourist enterprises located in countryside or urban areas ranging from city breaks, weekend breaks to national parks and other areas of outstanding natural beauty.

◆ Figure 1.14 shows the location of areas of outstanding scenic beauty throughout Britain which have become magnets for tourists and their location relative to major centres of population.

KEY

- Conurbations
- National Parks
- Areas of outstanding natural beauty

Scotland
A Gairloch
B Glenaffric
C Cairngorm
D Glencoe
E Loch Lomond

England and Wales
1 Northumberland
2 Lake District
3 Yorkshire
4 North Yorks. Moors
5 Peak District
6 Snowdonia
7 Brecon Beacons
8 Pembrokeshire Coast
9 Exmoor
10 Dartmoor

Figure 1.14 National parks, conurbations and motorways in the UK

Case Studies – Interactions

Key Idea 5

With reference to case studies of an upland glaciated area, an area of upland limestone and a coastal area you should be able to show knowledge of four interactions. These interactions are:

- social, economic and environmental impact of land use
- conflicts which can arise between land uses
- management strategies and solutions used to deal with land use issues
- the role of public and voluntary bodies in the management of land use strategies.

Social, economic and environmental impacts of land use

- Social impacts may involve the effects of different land use on the local population, local employment, housing and population trends.
- Economic impacts involve the effects of various land uses on the economy of the area, local economic activities such as industry, farming, tourism, local employment and whether these activities have improved the wealth of the area or created problems.
- Environmental impacts refers to the effects that various land uses have on the natural environment including the scenic value, wildlife habitats, levels of pollution and the overall quality of the environment.

Conflicts between land uses

There is great potential for conflict between various land uses in each of the study areas. Competition for land is perhaps the major cause of conflict between land users in the study areas. Figure 1.15 shows a range of competing land uses in rural areas.

In the examinations at both levels 1 and 2, you can be asked to describe the reasons why certain land uses are in conflict with each other. Usually you will be asked to select two or more from a given list or diagram. Alternatively you may be asked to look at an Ordnance Survey map and suggest land uses which may be in conflict and account for the nature of the conflict.

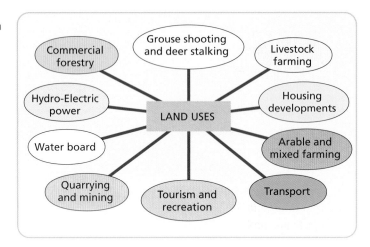

Figure 1.15 **Competition for land use**

◆ **Farming** requires substantial areas of land and all other land uses such as settlement, industry, forestry and reservoirs obviously limit the amount of land available. In rural areas most other land uses are in conflict with farming.

◆ Land use which produces any kind of pollution such as fumes, litter, chemicals or noise can have a detrimental effect on crops and livestock, so industry, settlements, communications and tourism will often be in direct conflict with agriculture.

◆ Farming can also produce its own chemical pollution which can affect wildlife (through use of pesticides and fertilisers) and visual pollution (through the use of poly-tunnels, for example), and this may impact on tourism.

◆ **Tourists** in any rural areas can seriously affect other land uses either directly or indirectly. Large numbers of tourists require certain facilities such as roads and other services.

◆ Large numbers of tourists can threaten farming in a variety of ways including soil erosion, leaving litter, pollution of rivers and lakes, traffic congestion, dogs worrying sheep, leaving gates open and trespassing on farmland.

◆ Forest areas may be endangered due to tourists and campers being careless with fire or damaging footpaths and forest walks with litter.

◆ All the study areas draw huge numbers of visitors, especially in the summer and this presents a further threat to the natural environment.

◆ **Traffic** issues such as road congestion and traffic fumes can affect natural vegetation and disturb the peace and quiet of the rural environment.

◆ **Industrial and settlement** land uses use up large areas of the countryside. Cities are always growing outwards and often expand into rural areas including small rural settlements.

◆ Apart from competition for land, some land uses can be in direct conflict with each other since their activities may directly affect or threaten the other. Agriculture and tourism are good examples of this.

You should be able to identify when and where conflicts exist, and also be able to describe the kinds of measures which can be taken to resolve them, either at local or national level.

◆ There are certain government agencies, at local or national level, whose task it is to monitor and wherever possible prevent or resolve major conflicts. These include National Park and Country Park Authorities.

Management strategies and solutions adopted to deal with identified land use issues

◆ National Park Authorities, Country Park Authorities and local pressure groups often ensure that the countryside is protected and that conflicts caused by competing land uses are resolved.

◆ Occasionally this can involve employing legislation to restrict access to certain areas or protective measures such as constructing footpaths to guard against erosion.

◆ Park authorities employ rangers to patrol the parks. The role of the rangers is to monitor potential problems and take action to resolve these problems.

◆ Planning legislation can be used to prohibit inappropriate land use such as housing or industry.

◆ Zoning of tourist areas can be encouraged by providing education centres for visitors and visitor centres.

◆ These strategies have been very successful in ensuring that tourism is sustainable and that the natural environment is protected and conserved.

The role of public and voluntary bodies in land use management issues

◆ In addition to National Park Authorities there are a number of public and voluntary bodies which have a major role in protecting and conserving the countryside of Great Britain.

◆ These include the National Trust, Countryside Agency, Scottish Natural Heritage (in Scotland), Campaign for the Protection of Rural England (in England), Royal Society for the Protection of Birds, Wildlife Trusts, Coastal Protection Agencies and a variety of local pressure groups, which take an active role in the process of protecting natural environment.

◆ Many of these organisations, such as the National Trust and the Royal Society for the Protection of Birds are funded through public donations.

◆ These public and voluntary bodies have been very successful in their efforts to maintain and conserve the natural environment in many areas. Their work has often led to public enquiries into various developments which could have a detrimental effect on the environment such as the building of nuclear and hydro-electric power stations, industrial estates, new motorways and similar projects.

Methods for the conservation, protection and management of scenic areas under threat

◆ Purchase of land known for its scenic beauty by voluntary and public bodies so they can control land use. This can help protect such areas from environmental misuse.

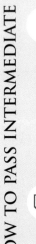

- Public awareness campaigns through sponsored education programmes, local newspapers and pamphlets and leaflets.
- Protests (such as public meetings and demonstrations) against measures which could damage the environment. These protests often target issues such as industrial and traffic pollution and congestion in rural areas, construction of new motorways through country parks, and blood sports such as foxhunting.

Three Sample Case Studies

Key Idea 6

In class you should have selected and studied several examples of areas within Britain with different landscapes. For the purpose of this revision text, three sample areas have been selected for brief case study.

These areas are the Lake District (upland glaciation) the Yorkshire Dales (upland limestone) and the Dorset and New Forest coastal area. These sample case studies may not be the same as those selected for your own study areas. However they provide a basis for answering questions on areas which you have studied.

Case Study 1 The Lake District (glaciated upland)

- Three main rock types make up the Lake District. These are igneous rocks such as granites, metamorphic rocks such as slate, and sedimentary rocks such as grits and limestone.
- This area contains England's best example of an upland glaciated area with a wide variety of features including pyramidal peaks, tarns, ribbon-shaped lakes, U-shaped valleys and hanging valleys.
- The most famous lakes include Windermere, Ullswater, Coniston Water and Thirlmere.

Land uses

Farming
- Due to the steepness of the slopes, the cold temperatures and high rainfall, which limit the growing season and affect soil fertility, the area is mostly unsuited to crop farming since it would be almost impossible to use machinery such as combine harvesters.
- The only type of farming which is feasible on a large scale is hill sheep farming, with cattle occasionally being raised on lower, less steep land.

Forestry
- Large plantations of coniferous forests are found in this area. This is suited to the steep slopes, poor soils and inhospitable climate. The trees also protect the slopes from soil erosion.

Industry
- Due to the lack of flat land, not much manufacturing has been attracted to the area. The main type of industry includes quarrying for granite and slate for roads and roofs.
- Limestone is also quarried for use in steel making elsewhere.
- The number of quarries operating has gradually been seriously reduced in recent years.

Water supply
- The Lakes have supplied Manchester with water for over 100 years despite being over 150 kilometres distant from the city.
- The Lakes are natural reservoirs in an area of high rainfall. It is much more economical to use these natural reservoirs than to build man-made reservoirs.
- The Lakes supply up to 30% of the water needs of this part of Britain.

Recreation, leisure and tourism
- The area is very attractive to tourists offering a variety of physical attractions such as the mountains and the lakes for activities such as hill walking, mountain climbing, adventure holidays, water sports, fishing and general sight seeing.
- Over 12 million people visit the Lake District each year and the number is increasing annually. 90% of visitors travel by car.
- 50% of visitors are from either regions very close to the Lake District such as Newcastle, Manchester and Leeds or from areas linked by the M6 such as Birmingham.
- Recent developments have included extensions to hotels and leisure complexes, timeshare complexes and marinas and mountain bike trails.
- The area was designated as a National Park with the main purpose of offering city dwellers a place to escape from the city and enjoy the benefits of a protected countryside.

Social, economic and environmental impact of the main land uses in the Lake District

Economic impacts
- In commercial terms, the land is very important to the local economy. The various land uses provide employment and income in a variety of ways.
- Many jobs have been provided in the area by tourism and in some parts 50% of the workforce is employed in the tourist industry.
- Increased tourism has brought more money into the area through tourist use of services, shops, hotels and boarding houses.
- The tourist industry has helped local employment opportunity with more jobs becoming available. This is especially important since tourist jobs have replaced jobs lost in agriculture.
- There are also many other businesses which rely partly on tourism for employment such as the building and catering industries.

Social impacts
- Socially, the land use activities have helped maintain the population of the area, whereas other parts of Britain with similar physical characteristics such as North West Scotland have become seriously depopulated.
- However, much of the existing housing is used for second homes for holiday makers. This makes it difficult for local people (especially young people) to afford housing, and many young people have to move out of the area. Housing is in short supply in the area and house prices have increased dramatically.
- This causes resentment among the local population due to the loss of trade to local businesses when second homes are left empty during the year.

Environmental impacts

◆ Efforts are made to protect the physical environment by National Park Authorities and other bodies. The environment has benefited from National Park status.

◆ Access has been made easier through the construction of motorways such as the M6, bringing many visitors from the south of Britain to the area.

◆ Land use has had a significant impact on the natural environment with increases in pollution, traffic congestion, footpath erosion and changes in the traditional rural character of many villages.

Main land use conflicts occurring within the Lake District

Due to competition for land, there is conflict between most of the major land uses. One major area of conflict exists between tourism and farming, which can be due to a range of factors.

◆ Increased traffic congestion, due to tractors holding up traffic and heavy use of small rural roads by tourist traffic.

◆ Damage caused by tourists in farming areas by litter, damage to fields, increased pollution, gates left open, animals worried by family pets. In response, some farmers have blocked access to public footpaths.

◆ There is also conflict between farming and other major land users, especially water boards who flood areas of valuable farmland and developers looking for land for industrial purposes.

◆ The increased use of lakes for water skiing, power boating, and jet skiing creates conflict between these activities and other lake users such as swimmers, sailors and anglers.

Main management strategies employed to deal with land use issues in the Lake District

◆ Legislation such as the Green Belt Act is enforced to protect and conserve the area from industrial and urban developments.

◆ Some strategies involve partnership between the National Park Authorities, the National Trust and the tourist and hotel industries to encourage sustainable tourism. This involves raising awareness of visitors, raising money to restore and conserve the landscape and ensuring that tourism and conservation work together to benefit the local community.

◆ Strict planning laws are observed to ensure that any development is both in character with the area and is environmentally sustainable. This means that new developments should not adversely affect the environment or economy or social character of the Lake District in the long term.

Role of public and voluntary bodies in the Lake District

◆ The National Park Authority has adopted many measures to protect and conserve the natural environment including the landscape, local vegetation and wildlife.

◆ These measures include those noted earlier such as zoning of activities, reducing traffic congestion, ensuring that a reasonable amount of new housing is sold only to local people and providing information and education centres for visitors.

◆ Approximately 25% of the land in the Lake District is owned and protected by the National Trust.

◆ The Trust provides footpaths for tourists, thus reducing the risk of erosion. It also maintains dry stone dykes and the habitats of wildlife in the area and therefore helps to reduce conflict between tourists and local farmers.

Questions and Answers ?

Question 1.6

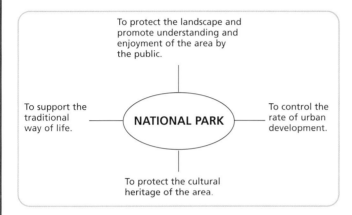

Figure Q1.6 Aims of the National Park

Study Figure Q1.6. Referring to a National Park you have studied, describe the strategies used to protect the environment of the area.   *(6 marks)*

Intermediate 2 2002 1ciii

Answer

In Loch Lomond in Scotland the environment has greatly been affected by tourism. The tourists have brought more noise pollution, air pollution and visual pollution (✓). The noise pollution comes from the many cars arriving at the loch but also from the jet skiers (✓). These jet skiers also disrupt the marine life in the loch by disrupting them with noise. Sometimes the jet skis leak oil or petrol killing fish and much of the marine life (✓). When tourists are walking through forests or farms they either drop litter or stray from the paths (✓), when doing this causing visual pollution as when they stray from the paths they wear down the rest of the beautiful scenery (✓). Trees have been cut down for more development.

The measures taken are that Loch Lomond has been made a national park so it is now properly managed (✓). The loch has been zoned to stop conflict between tourists but this also prevents the damage caused by say jet skiers spreading on the loch (✓). The trails are well monitored with rangers now being employed (✓) and more bins being provided for litter (✓).

Questions and Answers continued ➤

Questions *and* Answers *continued*

Comments and marks obtained

This answer provides a number of very valid points on both the problems caused and the measures taken to resolve these problems. Marks are obtained for reference to visual pollution, jet skiers, pollution of loch from oil spills, litter from tourists, footpath erosion and forestry destruction – a total of 6 marks for which the candidate gains 4.

Marks are also gained for reference to National Park, zoning, rangers employed to monitor path erosion and litter bins, gaining the candidate extra 2 marks. The candidate obtains a total of **6 marks out of 6** for the whole answer.

Answer

In the Peak District they have many ways to protect the environment. Because of the increasing amounts of footpath erosion, they have built durable paths to reduce the amount of damage to the hillside(✓). As there is large amounts of limestone, quarries would be dug, but the national park stop the quarries coming in and digging anywhere (✓). They limit the amount of quarries and where they can go. They stop farms increasing their field size and cutting down hedges (✓) as this is a cause of soil erosion and the hedges are an important habitat for wildlife.

Comments and marks obtained

The first mark in the answer is not obtained until the candidate mentions 'building the paths to reduce damage to the hillside'. The second mark is obtained for the statement referring to the National Park stopping quarries being built and located at random. The next sentence unfortunately repeats the action taken to limit the distribution of quarries and gains no further marks. A final mark is given for the reference to stopping farms from cutting down hedges which are important habitats for wildlife.

The answer is good enough for a basic pass, namely **3 out of a possible 5**, but does not contain sufficient detail for full marks.

Case Study 2 The Yorkshire Dales (upland limestone)

◆ The Yorkshire Dales are located on the eastern side of the Pennine Hills. They consist mainly of two rocks, carboniferous limestone and millstone grit.

◆ The area contains the largest area of upland limestone in the UK. All of the landscape features discussed in the section on upland limestone can be found in this area.

◆ Due to difficulties presented by the physical landscape and local climate there are restrictions on land use.

◆ Main land uses consist of hill sheep farming, quarrying, water storage, forestry, military training and moorland.

◆ Between them farming and moorland use about 97% of the land in the area. Forestry is the next most important land use with 2%.

Social, economic and environmental impact of the main land uses in the Yorkshire Dales

Economic impacts

◆ Farming is very restricted due to local geology, making it unsuited to arable farming. Hill sheep farming is the most common type found here.

◆ Quarrying for limestone is the main industry apart from tourism.

◆ With an annual income of over £6 million, the eight quarries of the area provide work for over 1000 employees.

Social impacts

◆ The area is popular for its underground features and attracts potholers and cavers.

◆ With over 2000 km of public rights of way paths, hill walking is a very popular activity in the area.

◆ The tourist industry provides a boost to the local economy by attracting an annual income in excess of £30 million and employment for over 1000 people.

Environmental impacts

◆ Forestry exists only in the millstone grit areas since trees do not grow well in the dry limestone areas.

◆ Due to the permeability of local rock water storage schemes are severely restricted.

◆ The area is not heavily populated and this makes it suitable for military training purposes since few people are disturbed by this land use.

◆ These activities do impact on the environment to some extent but the impact is not felt as much as in other parts of Britain where economic development poses much more of a threat such as in the Lake District.

Main land use conflicts within the Yorkshire Dales

As in the Lake District there is conflict between major land users such as hill sheep farmers, quarries, tourists, military training and housing.

◆ Limestone is in great demand for the building industry. However, quarries are often in conflict with other land users. This is due to the noise, traffic congestion caused by lorries used to transport the limestone and general pollution created.

◆ Many tourists are attracted to the area due to the scenery. Hill walking is very popular but has potential for eroding the landscape.

◆ Traffic congestion caused by increased tourism is also a source of conflict between the local population and the visitors to the park.

Main management strategies employed deal with land use issues in the Yorkshire Dales

◆ The area was designated a National Park in 1954. The National Park Authority employs many of the same measures as in the Lake District to protect and conserve the natural environment.

- Planning permission can be refused to prevent companies opening new quarries and the National Park Authority can insist that quarries be screened from public view.
- There is planning restriction on the building of new developments such as housing, industry and other developments which may adversely affect the human and natural environments.
- Farmland is protected from the impact of tourism and other activities by the provision of footpaths and restricted access to certain areas.
- Greenbelt legislation is enforced to limit developments such as new housing and industrial land use.

Role of public and voluntary bodies in the Yorkshire Dales in dealing with land use issues

- Laws limiting developments, which could spoil the natural environment, have helped to maintain the scenic value of the area. This benefits local farmers, residents and tourists who visit the area.
- Public and voluntary bodies active in the area include the National Trust.
- An educational charity called the Yorkshire Dales Society was also set up to promote the conservation of the Dales scenery and the ways of life of its people. This society organizes walks and lectures and makes great efforts to limit quarry development in the area. The society tries to promote sustainable development by encouraging the provision of alternative employment to industries, which cause environmental damage.

Case Study 3 The Dorset and New Forest coastal area

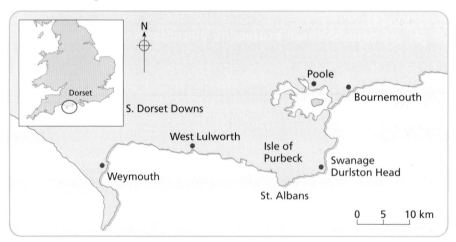

Figure 1.16 Dorset Coastal Area

- The Dorset Coastal area extends between the mouths of the Lymington and Beaulieu Rivers. There is a wide variety of land use present including farming, industry, forestry, tourism, recreation and leisure.
- The estuaries of the Lymington and Beaulieu rivers are centres for sailing, boat building and repairs.

◆ In the area of Southampton Water much of the area is developed for major industry such as the Esso oil refinery and petrochemicals complex, housing developments and power stations.

◆ Coastal marshes in the area are used for nature conservation despite the risks from water pollution caused by industrial waste and domestic sewage.

◆ These activities have a major impact socially, economically and environmentally on this area of coastline.

Social, economic and environmental impact of the main land uses

◆ Social impact is felt in terms of the area being a magnet for population. The area is very attractive to tourists, and as a place to live, work and retire.

◆ 17 million tourists are attracted to the area each year.

◆ There are over 200 000 educational visitors catered for every year.

◆ Economically the area benefits from the wide range of economic activities including tourism, industry, forestry and farming. These activities generate a great deal of money for the people who live and work in the area.

◆ Environmental effects of these demands from tourism, port and ferry services and increased traffic congestion include, footpath erosion, demand for land for car parks and other amenities, threats to wildlife habitats, danger of marine pollution, conflicts between fishing, water sports and marine archaeology and the presence of the UK's largest onshore oilfield in the area at Wytch Farm.

Main land use conflicts within the Dorset coastal area

All of the land use activities combine to threaten the natural environment and to change the natural balance and ecological diversity of the area. In addition the coastline is under threat from natural forces such as waves, currents and groundwater movement.

Management strategies to deal with land use issues

A document outlining the strategy for sustaining and improving the quality of the environment called 'New Forest 2000' was published in 1990.

◆ This strategy included measures to reduce pollution levels, protect the scenic beauty, improve the appearance of the coast, to maintain the economy, to protect the coastline, to educate the public and to conserve features of historic and archaeological interest.

◆ The coastline needs to be managed to sustain human activities from the threat of marine erosion, to preserve coasts for conservation reasons and to preserve them from development.

◆ Responsibility for managing coasts generally lies with three agencies: the Environment Agency; the Department of the Environment, Food and Rural Affairs (DEFRA); and the local district councils.

◆ Coastal defence strategies include measures such as seawalls, the use of large irregular rocks, gabions (wire baskets filled with rubble), groynes and embankments. There have also been attempts to protect coastlines from flooding using dykes and flooding walls.

◆ Each of these measures has advantages and disadvantages, not the least of which is cost. Some are relatively cheap whilst others such as seawalls can be very expensive to build and maintain.

◆ Nature Reserves have been created to protect wildlife and two country parks have been created to encourage sustainable tourism.

◆ Public authorities, which operate the nature reserves and the country parks, manage these strategies, and the area has achieved the status if not quite the title of a National Park.

These efforts in the New Forest Area have met with considerable success as have those in the other study areas discussed.

Role of public and voluntary bodies in the Dorset coastal area in dealing with land use issues

Public and voluntary bodies, other than those associated with Nature Reserves, are involved in efforts to resolve land use issues include coastal management authorities. These include:

◆ the designation of Heritage Coast Status for the area;

◆ designation of selected areas as Sites of Scientific Interest and Special Marine Conservation Areas.

All of these bodies make great effort to ensure that the natural environment of this area is protected as much as possible from the pressures and threats outlined above.

Questions and Answers

Question 1.7

(i)   With reference to an area in the British Isles you have studied, describe the environmental problems created by increased tourism.

(ii) Explain the measure taken to minimise the long-term impact of tourism.
(5 marks)

Intermediate 2 2003 2ci

Answer

(i) Dorset – Problems created were seasonal unemployment because tourists tend to only come in the summer months. Honey pot problems occurred between the fishermen and water sports because water sports scared fish away (✓). Increased tourists meant increased traffic congestion (✓) – so it would annoy locals getting to work etc., Increased traffic brought more noise and air pollution (✓) and the more tourists meant increased litter (✓).

(ii) Measures taken to minimize this long term impact is park and ride schemes being set up and more car parks (✓), so that there would be less traffic congestion. Zoning took place giving fishermen and water sports their own parts to go to (✓).

Questions and Answers continued ➤

Questions and Answers continued

Comments and marks obtained

The opening statement does not refer to conflict and does not merit any marks. The next point refers to conflict between fishermen and water sports and gains a mark. References to tourists affecting local population gain a further three marks.

The comments on the measures taken 'park and ride schemes/car parks' and 'zoning in water areas' gains a further two marks. The answer has sufficient valid points to obtain a total of **5 marks out of 5**.

Glossary *Physical Environments*

Abrasion: The process by which rocks within ice sheets and rivers scrape and erode the land over which they pass.

Alluvium: Material deposited by a river usually over its flood-plain.

Arête: A narrow ridge between two corries, formed as the corries are formed on two adjacent sides of a mountain.

Attrition: The process which occurs as rocks in a river wear away by constantly rubbing together.

Braiding: The process by which rivers divide into separate channels through material being deposited by the river mid stream.

Corrie: An arm chair shaped hollow on the side of a mountain which was formed by ice filling a hollow and eroding the side of the mountain by abrasion and plucking and rotational movement at the base of the hollow. When the glacier melts sometimes a lake or loch is left called a corrie loch, lake or tarn.

Country Park: An area in the countryside surrounding a town or city which has been set aside for people to visit as a park.

Countryside Agency: A government organization which monitors and protects countryside areas from harmful development.

Drift: Material deposited by a glacier and is made up of two main parts known as *till*, which is deposited under the glacier, and *outwash* which is formed by melt water streams carrying particles of material from the debris under the glacier.

Distributary: A branch of a river flowing out from the main river.

Drumlin: An oval shaped hill formed from deposits from a glacier.

Erosion: The process by which rocks and landscapes are worn away by agents such as moving ice, wind, flowing water and sea and wave action.

Glossary *Physical Environments continued*

Erratics: Rocks or boulders which have been moved by ice sheets from their original location and left in other parts of the country during the Ice Age.

Eskers: Long ridges of sand and gravel deposited by rivers which flowed under ice sheets.

Estuary: The mouth of a river where the river meets the sea.

Flood-plain: The area on either side of a river, usually in its middle or lower stage which is formed from material carried by the river and deposited when a river overflows during a period of flooding. Flood plains can vary greatly in size depending on the stage of the river and the deposited material is called *alluvium*.

Forestry Commission: The organization responsible for planting and maintaining forests throughout the UK.

Freeze thaw action: An erosion process which occurs when water trapped in cracks in rocks alternately freezes and thaws causing the rock to break up.

Frost shattering: Similar to freeze thaw, caused when water turns to ice and then expands as it melts, putting pressure on the rock eventually causing it to shatter. Material from this may fall to the bottom of slopes and gather as rock debris known as *scree*.

Glacier: A large mass of moving ice that changes the shape of the land over which it is passing.

Grykes: Deep joints or fissures on the top of limestone plateaus formed through the chemical reaction of rainwater and limestone. When deep enough they leave the surface area to form large blocks called *clints* giving a pavement like appearance.

Hanging valley: A valley on the slopes of a U-shaped valley, usually a tributary of the main valley.

Land use: The ways in which humans make use of the physical landscape e.g. forestry, farming, industry or settlement.

Land use conflict: Conflict occurs when different activities compete with each other to make different use of the same land such as farming and tourism.

Mass movement: The process by which rocks move under gravity.

Meanders: Bends formed in the middle or lower courses of rivers.

Moraine: Material deposited by glaciers. Different types include *end-moraines* (or terminal moraines) formed at the front of the glacier as it melts, *lateral moraines* formed at the sides of glaciers and *medial moraines* formed in the middle of glaciers or at the edges of where two glaciers meet.

National Park Authority: Organization which looks after areas which have been set aside throughout Britain for the recreation and enjoyment of the public. Their aims also include protecting these areas of outstanding scenic beauty.

Ox-bow lakes: Lakes formed from former meanders when the loop of the meander is cut off from the main channel by a build-up of deposits from the river.

Glossary *Physical Environments continued*

Plucking: The process by which moving ice tears rocks from the surface over which it moves.

River cliff: The steep bank of a river formed as the river undercuts or erodes the outside bank.

Stalactites and stalagmites: Features formed underground in limestone areas where material in solution is deposited on the floor or drips from the ceiling of underground caves. The moisture evaporates leaving small deposits of calcium carbonate which gradually build up to form the stalactites from the roof and stalagmites from the floor. These can eventually meet to form pillars.

Source: The point at which a river begins.

Till: Material deposited beneath a glacier. It consists mainly of boulder clay.

Transportation: The process by which rock particles are carried by rivers or glaciers or wind.

Tributary: A small river which joins with a larger river.

Truncated spur: A sloping mountain ridge which is cut away or eroded by a glacier to form a blunt-ended feature. Often found beside hanging valleys.

U-shaped valley: A valley with very high steep sides and a wide flat bottom formed by a glacier flowing through the original valley. It is usually occupied by a small river which is known as a *misfit stream* since it was not the original river flowing through the valley.

V-shaped valley: A narrow valley which is typical of a valley in the upper or youthful stage of development.

Waterfall: A steep break in the course of a valley often associated with changes in rock type along the course of a river.

Weathering: The process by which rocks are worn away through physical action such as flowing water, wind or a chemical reaction between rocks and rainfall which may have become acidic.

HUMAN ENVIRONMENTS

World Population Distribution

Key Idea 1

You should be able to describe and explain world population distribution.

Introduction

Population is not spread evenly throughout the world. Population density is often used as an indicator of the distribution of population throughout the world. Population density is usually expressed as the number of people per square kilometre. The pattern of world population distribution is shown in Figure 2.1.

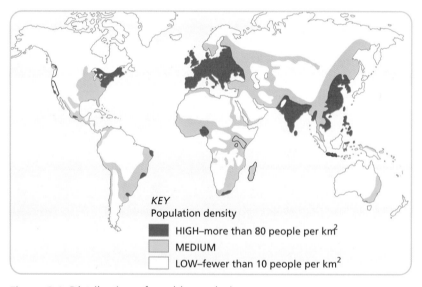

KEY
Population density

■ HIGH–more than 80 people per km^2

▨ MEDIUM

□ LOW–fewer than 10 people per km^2

Figure 2.1 Distribution of world population

> ## Key Point 1
>
> When asked to explain patterns of distribution you should:
>
> ◆ describe where the main areas of population occur or perhaps where the density of population is high, medium or low
>
> ◆ explain world population distribution by referring to the comparative influence of physical and human factors
>
> ◆ understand that factors which attract and repel population are called positive and negative factors respectively.

Physical factors which influence population distribution and density include relief, climate, natural resources and employment opportunities.

Relief

Positive factors

◆ Flat land for building settlements.

◆ A good supply of water from rivers and valleys.

◆ Easy access such as location in a valley or on a coastal plain.

Negative factors

◆ Situations which are inaccessible such as mountains, jungle and deserts.

◆ Situations which are remote.

◆ Situations which are difficult to build on such as steep mountain slopes.

Climate

Positive factors

◆ Climates which are suitable to live and work in.

◆ Temperate climates which have mild to warm temperatures throughout the year and moderate or adequate amounts of rainfall for water supplies.

◆ Climates which are suitable for growing crops.

◆ Climates which are suitable for the development of tourism such as Mediterranean climates which are warm and dry in summer, mild and wet in winter.

Negative factors

◆ Areas where the climate is inhospitable such as being too hot, too cold, too wet or too dry.

◆ These include climates such as hot deserts, tundra, Arctic and Equatorial and tropical climates.

Natural resources are a source or potential source of wealth. If these are accessible, large numbers of people are usually attracted to them.

Positive factors

◆ Good fertile soil and flat land for farming. Areas where labour is the main input for farming. Such areas include those where the system of farming is based on intensive peasant farming. Examples of these include India and the countries of South East Asia.

◆ Mineral resources such as oil, coal, ores, and precious metals such as gold and silver. These resources attract both primary and manufacturing industry and consequently large numbers of people to work in these industries.

Negative factors

◆ Areas which lack natural resources do not offer much opportunity for human development and settlement. Such areas are typified by low population density.

Employment opportunities

◆ Due to a combination of equitable climate conditions, suitable relief and the presence of natural resources, many areas of the world are highly suitable for industrial and agricultural development.

◆ These developments require large numbers of people for labour.

◆ In Economically More Developed Countries (EMDCs) most of the workers find employment in industries in towns and cities.

◆ In Economically Less Developed Countries (ELDCs) the majority of workers are usually employed in some form of farming.

◆ As agriculture and industry develop, more jobs are created attracting more people from other areas.

◆ In EMDCs the majority of industrial jobs are in high-tech manufacturing and the service sector.

◆ In ELDCs most industrial jobs are in primary industries such as mineral extraction (mining and quarrying), forestry, fishing and farming.

Questions *and* Answers (?)

Question 2.1

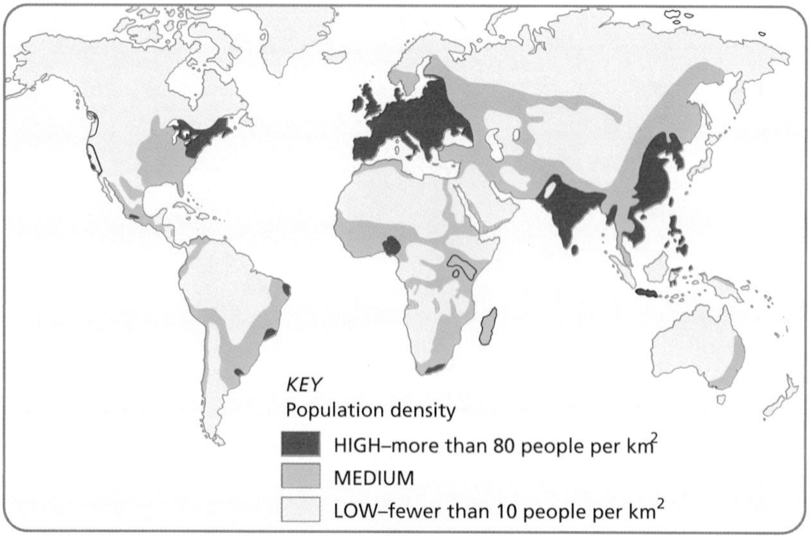

KEY
Population density

▣ HIGH–more than 80 people per km^2
▨ MEDIUM
☐ LOW–fewer than 10 people per km^2

Figure Q2.1 World population density

Questions and *Answers continued* ➢

Questions *and* **Answers** *continued*

Look at Figure Q2.1. Explain why some parts of the world are more densely populated than others. Refer to both physical and human factors in your answer. *(4 marks)*

Intermediate 2

Answer

Some parts of the world are more densely populated than others because some areas have plenty of resources (✓) and good flat land for building on (✓). Also some countries are richer than others attracting more people to them. Good climates also attract people (✓) and that is why some areas are more densely populated than others.

Comments and marks obtained

The references to resources and flat land for building on gain a mark each. However the reference to richer countries does not merit a further mark since many poor areas and countries in the world have high density populations such as India and South East Asia. The comment in the last sentence about climate merits the third mark. This is a good answer but will gain only **3 out of the 4 marks** available.

Population Structures

Key Idea 2

You should be able to describe and explain population structures including urban and rural structures. Population structures are shown by diagrams called population pyramids.

Population structure

◆ The structure of the population of a given country is defined in terms of age and sex distribution. Males and females are sub-divided into different age groups e.g. 0 to 4, 5 to 9 and so on up to the 80+ category.

◆ Data on these characteristics are plotted on a graph called a *population pyramid* shown in Figure 2.2.

◆ Structures can indicate variations in levels of development.

◆ Analysis of population structure graphs (age/sex pyramids) can reveal patterns of birth and death rates and an estimation of life expectancy.

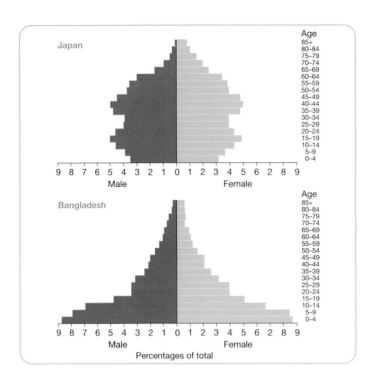

Figure 2.2 Population structures of Japan and Bangladesh

Population structure in an Economically More Developed Country (EMDC)

The population structure for Japan (an EMDC) shows a typical population structure for an EMDC.

◆ Birth and death rates are fairly low, and people have a fairly high life expectancy,

◆ The number of people of both sexes in the groups beyond 60+ is a relatively high proportion of the total population.

◆ The structure is well balanced with the number of males and females evenly distributed throughout the different age groups.

◆ The shape of the pyramid shows a very gradual tapering off towards the upper age groups.

◆ There is no imbalance between the lower groups between 0–15 and the groups between 15–60.

◆ In effect the base is neither significantly wider - indicating a high birth rate, or significantly narrower – indicating quite a low birth rate.

◆ The very gradual narrowing of the groups towards the upper end indicates quite a low death rate in both males and females.

Population structure in an Economically Less Developed Country (ELDC)

The population pyramid for Bangladesh (an ELDC) shows a typical population structure for an ELDC.

◆ This graph has a wide base for both males and females up to the 10–14 age group. This indicates a high birth rate which is typical of an ELDC.

◆ The middle section of the graph becomes much narrower than the graph for Japan.

◆ The upper part in the 50–60+ age groups becomes very narrow suggesting a low life expectancy rate.

Differences in urban and rural structures

There are wide variations in the percentages of urban and rural populations in many countries throughout the world.

◆ In ELDCs the percentage of rural population is usually higher than urban population.

◆ In EMDCs the vast majority of the population live in urban areas. In Europe and North America more than 85% of the population is urbanised.

◆ In ELDCs the balance between rural and urban population is changing as people leave the countryside and migrate to urban areas.

◆ When this migration takes place, the structure of the remaining population may be left imbalanced with a high proportion of older people.

◆ During the late 19th and early 20th centuries millions of people left rural areas in Europe to seek new lives abroad. This was true of countries like Italy, Ireland, Germany and Russia. Many people emigrated from these countries to the USA for social, economic and political reasons.

World Population Change

Key Idea 3

You should be able to describe and explain world population change. You should do this by referring to the following measures of population.

◆ **Birth rates (crude)** This figure indicates the number of people per thousand of the population, born in any given year. Since this is the basic measure it is termed 'crude'.

◆ **Death rates (crude)** This indicates the number of people per thousand of the population who die in any given year.

◆ **Natural growth rate** Subtracting death rates from birth rates gives a basic indication of the number by which the population is increasing each year per thousand of the population.

◆ **Average life expectancy** This is a figure which indicates the average number of years a person can expect to live within any given country. Even within one country, average life expectancies are usually different for men and women.

◆ **Infant mortality rate** This rate indicates the number of deaths per thousand of the population in a year in any country of children under the age of one year.

Key Point 2

You should be able to describe and explain factors affecting population change in different parts of the world including influences on birth, death and infant mortality rates.

Factors leading to high birth rates include:

◆ the need to use children as labour to increase family income
◆ religious beliefs or lack of education preventing the use of artificial methods of birth control
◆ traditional and cultural reasons for large families, particular in cultures where there is an expectation that large families will support parents in old age
◆ possible insurance against high infant mortality rates.

Factors leading to low birth rates include:

◆ people marrying later for economic reasons
◆ widespread availability, education and use of birth control and contraception methods
◆ women in developed countries putting career before having children.

Factors leading to high death rates include:

◆ widespread poverty, poor diet and malnutrition
◆ widespread disease due to poor health care, poor hygiene and sanitation, poor access to clean water supply
◆ lack of basic health care, medical drugs, hospitals, clinics and doctors
◆ unhealthy environments
◆ natural disasters such as drought, floods, and earthquakes.

Factors leading to low death rates include:

◆ economic prosperity
◆ availability of good health care, hospitals, doctors, medicines
◆ plentiful supply of food and well balanced diets
◆ availability of good housing, sanitation and clean water
◆ safe natural environments
◆ high standards of education and health education.

Other factors affecting population growth rates include:

◆ high or low rates of immigration and emigration
◆ levels of economic development in a particular country.

Factors affecting infant mortality rates include:

◆ standards of health care
◆ cultural and religious influences on the provision of birth control measures
◆ quality of food supply and variation in diets
◆ standards of housing, water supply and sanitation

◆ The occurrence of natural disasters such as tropical storms, earthquakes, volcanic eruptions, floods and droughts, which can affect food supply and often lead to widespread famine which greatly impacts on the very young.

Key Point 3

You should know the main factors which influence and create variations in life expectancy rates throughout the world. Life expectancy is the average age to which people can expect to live in any given country or region. A country's life expectancy rate depends on a combination of several factors including:

◆ diet, lifestyle, level of income, employment status, quality of living accommodation

◆ incidence of contagious and infectious diseases

◆ levels of health care available

◆ environmental factors such as climate (which can have impacts such as the proximity to breeding grounds of disease carriers such as mosquitoes)

◆ environmental hazards such as earthquakes, floods and droughts

◆ levels of pollution.

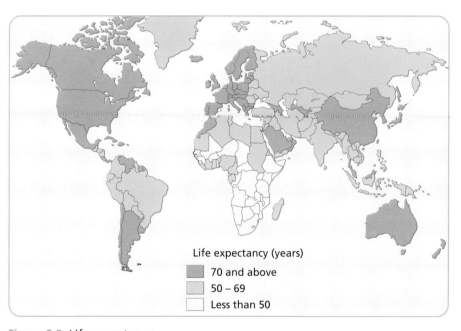

Figure 2.3 Life expectancy

As a result of combinations of all the factors mentioned here, economically less developed countries tend to have significantly lower life expectancy rates than economically more developed countries. Life expectancy is closely tied to levels of development which can vary greatly even within that group of countries described as economically less developed.

Key Point 4

You should be able to discuss reasons for the occurrence of high infant mortality rates in some parts of the world. You should also know about measures taken and their effectiveness in reducing high infant mortality rates.

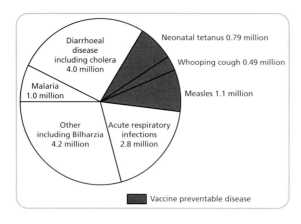

Figure 2.4 **Causes of deaths of children under 5 in developing countries**

Factors causing high rates of infant mortality

◆ Lack of food causing starvation and malnutrition due to famine.

◆ Poor medical provision including insufficient doctors and other trained medical staff and a limited supply of vaccinations and treatment for various diseases.

◆ Lack of clean water supplies resulting in the transmission of many water borne diseases such as cholera, schistosomiasis, malaria, typhoid and diahorrea resulting in infant deaths.

◆ Inadequate housing lacking basic amenities such as cooking facilities, water and electricity supplies and poor sanitation.

Measures used to reduce high infant mortality rates

◆ Education programmes on the use of birth control which reduces the size of families giving children a better chance of survival.

◆ More effective hygiene control being taught (such as the use of tablets to sterilise drinking water and boiling water to kill germs).

◆ The provision of additional medical facilities especially through primary health care schemes.

◆ Aid programmes to provide food and medical supplies for children.

◆ Assistance from local government authorities to provide and encourage self help schemes to improve living conditions for poverty stricken families.

◆ Government investment in agricultural output schemes to increase food production.

Key Point 5

You should be able to describe and explain how population growth rates and structures have changed over time. In the external exam, you may be asked to refer to the Demographic Transition Model, shown in Figure 2.5.

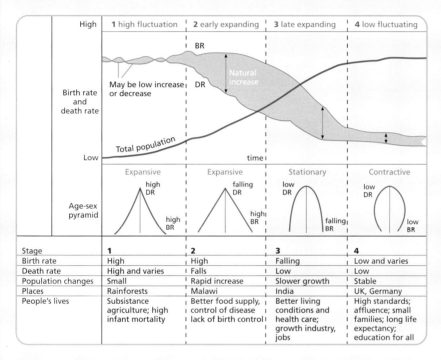

Stage	1	2	3	4
Birth rate	High	High	Falling	Low and varies
Death rate	High and varies	Falls	Low	Low
Population changes	Small	Rapid increase	Slower growth	Stable
Places	Rainforests	Malawi	India	UK, Germany
People's lives	Subsistance agriculture; high infant mortality	Better food supply, control of disease; lack of birth control	Better living conditions and health care; growth industry, jobs	High standards; affluence; small families; long life expectancy; education for all

Figure 2.5 The Demographic Transition Model

The Demographic Transition Model is a model of the ways in which population changes over periods of time at different stages of economic development. It shows four separate stages of population growth, and uses line graphs to indicate changes in birth, death and growth rates over time. It is called a transition model because it shows the way the population of a country changes as it moves through different stages of economic development.

The Demographic Transition Model summarises patterns of population growth in various countries throughout the world and is based on the relationship between births and deaths. You should be able to explain what is happening in each of the four stages of the model and give reasons why this is occurring. The changing population structures are indicated by the population pyramids shown at each stage of the model.

Stage 1
◆ Both birth and death rates are high and therefore population growth is low.
◆ Britain was in this stage about 250 years ago at just about the start of the period known as the Industrial Revolution. At this time industries were developing in towns and cities and attracted people from rural areas.

◆ Large families were often dependent on the income from children working in factories. Living conditions were poor with the majority of people living in overcrowded tenements with poor sanitation. Many died from diseases which today would be regarded as fairly minor.

◆ At the present there are several countries throughout the world where birth and death rates are both high. These are generally less developed countries such as Ethiopia.

◆ War, famine and droughts mean that the death rate is high and life expectancy is low. People continue to have large families despite the comparatively high infant mortality rates.

Stage 2

◆ The birth rate remains high whilst the death rate gradually falls, so the population growth rate increased.

◆ Britain was in this stage at the beginning of the 20th century. Advances in medicine, better diets and improvements in living conditions such as improved sanitation caused the death rate to begin to fall.

◆ Countries such as Mexico are at this stage today.

◆ Birth rates remain high due to the influence of factors such as religion forbidding birth control, lack of education and social reasons (such as the need for large families to care for parents as they get older).

◆ With the introduction of foreign aid programmes in developing countries bringing improved medical care and improvements in agriculture the death rate is gradually falling leading to rapid increase in population growth.

Stage 3

◆ The death rate continues to fall and the birth rate also begins to fall but at a much slower rate. Population continues to grow.

◆ Through long-term aid projects provided through the United Nations agencies many less developed countries are beginning to reduce both their birth and death rates. This includes countries in Asia, Africa and South America.

◆ Medical assistance is provided in the fight against various diseases such as cholera, typhoid, malaria and disease caused by hunger and malnutrition.

Stage 4

◆ Death rates continue to fall in more developed countries of the world, with advances in medical science and new drugs becoming available to treat major diseases such as cancer and heart disease.

◆ Birth rates also continue to fall due to changes in lifestyle with artificial contraception being widely available and widely used. Fewer women stay at home to look after large families.

◆ Birth rates and death rates are almost equal and as a result population growth is very small.

◆ Countries in this stage may find that their population consists increasingly of people in older age groups.

> ## Key Point 6
>
> You should be able to show knowledge and understanding of the implications of population change in EMDCs and ELDCs. Reasons for this pattern of change include:
> - widespread use of artificial birth control
> - changes in the status of women, with many women having jobs and careers in preference to marrying young and starting a family
> - changed attitudes of younger people towards marriage and size of families
> - opportunities for improving the standard of living through having fewer children.

Implications of falling birth rates in EMDCs

- Under-population, which occurs where the birth rates and death rates are very low and are almost the same. Population growth is very slow and in some cases is decreasing. When this happens the population structure becomes imbalanced.
- Fewer young people reduces the size of the available work force.
- More older people need to be cared for due to increased life expectancy.
- With fewer young people and an ever-increasing number of older people there is greater pressure on the economically active age groups to support the dependent groups.
- There is an increasing need for medical care (hospitals, clinics, medical staff and drugs) and special accommodation for the ageing population. More government money is needed to provide this. There is an increased need for more leisure activities for the elderly.
- Taxes on the working population need to increase to pay for the health and social services costs of the ageing population.
- The retirement age is often increased to save money on pensions.
- Government funds may be diverted from other areas of the economy such as education, transport, defence and so on.

Implications of rising birth rates and falling death rates in ELDCs

- Overpopulation, which is said to exist whenever a reduction in the existing population would result in an improvement in the quality of life for the remaining population.
- Insufficient food to meet the demands of the growing population.
- Inadequate housing for the population, particularly in cities and towns. This severely reduces the quality of life for a large proportion of the population.
- Vast numbers of people in less developed countries are forced to live in very poor accommodation such as slums, squatter areas and shanty towns.
- These areas often lack the basic facilities of sewage, electricity and water supply. As a result diseases such as typhoid and cholera become widespread. There is often an increase in infant mortality rates as a result of disease outbreaks.
- High unemployment which occurs since there are far too many people for the jobs available. As a result there will be widespread poverty. This is made worse by the lack of government financial aid.

◆ Insufficient services such as health centres, hospitals, doctors, schools, and colleges due to increasing demand from the growing population.

◆ This creates problems of poor health standards and poor education standards. Literacy rates (the percentage of the population which can read and write) are usually very low, often less than 50%.

◆ Increased migration from poorer to better off areas.

Key Point 7

You should be familiar with a variety of population graphs and be able to interpret them.

Geographical methods and techniques (Population graphs)

Analysing different types of graphs is important in the study of population. For example the analysis of the population model by reference to the different patterns of birth rates and death rates throughout the four different stages could be used to assess your geographical skills in this topic area.

This would involve being able to take each stage in turn and provide an analysis of whether the rates are rising or falling, how fast or slowly this is happening and what the net effect would be on population growth.

Note that you could be expected to explain why these changes are occurring as well as describing what is happening. In the external examination this could be combined with a question requiring you to explain the patterns in the different stages.

Alternatively, you could be asked to comment on the effects of these changes on the population. To answer such questions you would have to draw on your knowledge of population trends, patterns and the factors responsible. You should be able to relate these to case studies of countries you have studied.

Interpretation of population data

You will need to be able to use population data to explain trends. Questions on these kinds of diagrams may ask you to examine the graphs, tables or maps and draw conclusions.

◆ For example, you may be given a table showing various measurements of population for two or more countries. This may contain details on life expectancy, birth and death rates, infant mortality rates and other indicators such as medical provision.

◆ You may be asked to explain differences between the countries in areas such as life expectancy. You can do this by first describing the differences followed by reference to other information to support your conclusions.

◆ Similarly if you are asked to use diagrams such as population pyramids to explain differences between countries, you should be able to identify patterns of birth and death rates and life expectancy rates and to match them appropriately to countries which are either Developed (EMDCs) or Developing (ELDCs).

◆ Your knowledge of factors which contribute to levels of development should help you to explain both the structures and the reasons for them.

Questions and Answers

Question 2.2

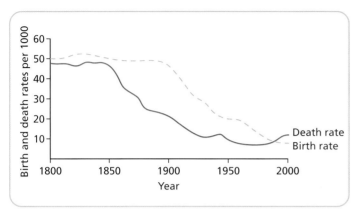

Figure Q2.2 Demographic changes in Sweden

Look at Figure Q2.2. Describe in detail the changes which have taken place in birth and death rates in Sweden since 1800.   *(3 marks)*

Intermediate 2 2002 4a

Answer

From the graph the death rate in 1800 was 47 per 1000 where the birth rate was 50 per 1000. The rates stayed the same until 1850 (✓) when there became a gradual decrease of birth rate and sudden decrease in death rate (✓) as in 1900 the death rate was 20 per 1000 and the birth rate was about 48 per 1000, where in 1900 there was a sudden decrease in birth rate (✓). In 1950 the death rate reached its lowest of 10 per 1000 (✓) and the birth rate was still decreasing at 20 per 1000 (✓). In 2000 the death rate has increased again gradually at 15 per 1000 (✓) but the birth rate has decreased to 8 per 1000 (✓).

Comments and marks obtained

This answer makes good use of the information provided on the diagram. The answer goes through the various stages of the diagram making a series of 6 valid points, quoting correct details from the diagram on changing birth and death rates, noting increases and decreases. The answer merits a full **6 marks out of 6**.

HOW TO PASS INTERMEDIATE 2 GEOGRAPHY

Key Point 8

You should know the reasons for effects and world patterns of immigration/emigration.

Migration patterns

◆ Immigration is the permanent inward movement of people from other parts of the world to a particular country. This normally results in an increase in population.

◆ Emigration is the permanent outward movement of population from a country to another part of the world. This would normally cause population size to decrease.

Figure 2.6 illustrates world patterns of migration.

Figure 2.6 **Patterns of world migration**

Reasons for migration

◆ Forced migration to provide cheap labour, such as the African slave trade of the 17th and 18th centuries. In this time up to 10 million Africans were forcibly moved from Africa to other parts of the world.

◆ Displaced migration as a result of war. In the First and Second World Wars large masses of people were displaced from their homelands to other countries.

◆ Economic migration, when people migrate from one area to another in the hope that they might improve their standard of living and general lifestyle. People often move from areas with high unemployment, low wages, poor housing, poor health and educational facilities and political and/or religious persecution. They move in the hope of improving on these conditions.

◆ Conditions which cause emigration are known as **push factors** and those resulting in immigration are termed **pull factors**.

- When people move from countryside to urban areas (towns and cities) these factors are called '**rural push and urban pull**'.
- These factors are illustrated in Figure 2.7, showing the Developing World Migration Model.

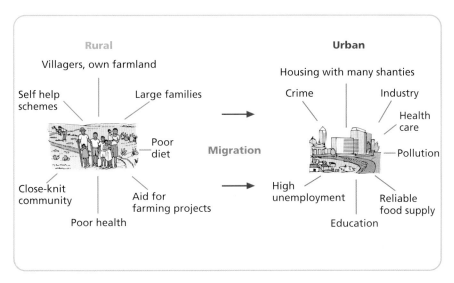

Figure 2.7 The Developing World Migration Model

Figure 2.7 shows the pattern of movement often found in less developed countries. A large portion of the population of such countries live and work in rural/agricultural areas. This is typical of countries like India, Brazil and countries of South East Asia.

Effects of migration

- Immigrants moving from rural to urban areas often end up living in shanty towns within and on the outskirts of cities. Many find themselves living in conditions which are worse than those which they left behind. They may be left disappointed and disillusioned.
- Shanty towns have the poorest living conditions including poor housing, lack of clean water supplies, no electricity and often the most basic systems of sanitation.
- Diseases such as cholera and typhoid are common in these areas. These diseases kill the weakest members of the community namely the poorest, the old and the very young.
- Immigration of different ethnic groups into countries can lead to problems of integration, racial tension and cultural differences. Immigrants are often targets for abuse and discrimination in terms of jobs and housing. Ghettos may develop in inner city areas involving groups of immigrants.
- Immigrants may be prepared to take lower paid jobs and incur the anger of the existing population for doing so. Some countries pass legislation both to limit immigration and protect new immigrants.
- Immigrants may also improve the cultural diversity and provide additional labour, especially in countries where the population is in decline due to falling birth rates.
- Emigration may leave communities without an adequate workforce or with population imbalance, especially in rural areas.

Questions and Answers (?)

Question 2.3

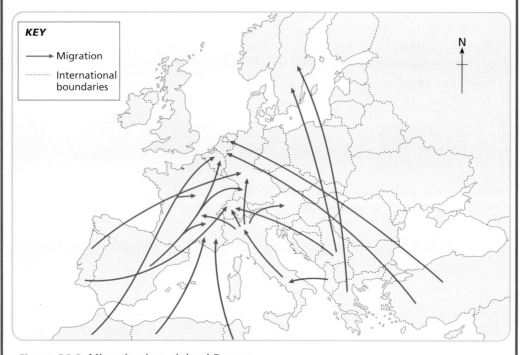

KEY

→ Migration

········· International boundaries

N

Figure Q2.3 Migration in mainland Europe

Study Figure Q2.3. Describe the effects of migration on any named mainland European country you have studied. *(5 marks)*

Intermediate 2 2003 4c

Answer

The effects of migration of the Turks to West Germany were that the dependant population were left behind. The country receiving migrants had some advantages as the need in low cost labour was reduced (✓) since the migrants would be happy to do low cost labour e.g. working in factories which was a good effect for the factory owners (✓) as they were not paying the amount of wages that they were meant to. (✓) The country losing the people had many effects on them as they were losing their active population and there were jobs without people (✓). The country receiving the migrants had an increasing population as the migrants were calling their families over to live with them. This resulted in more schools needing to be built and more houses (✓).

Questions and **Answers** *continued* ➤

Questions and Answers continued

Comments and marks obtained

The answer correctly identifies several advantages to the migrants and the receiving country such as, 'low cost labour' for three marks. Further marks are obtained for references to the impact of migration on the losing country – loss of active population – and the final reference for the receiving country needing more schools/houses gains the last mark. The answer merits a total of **5 marks out of 5**.

Question 2.4

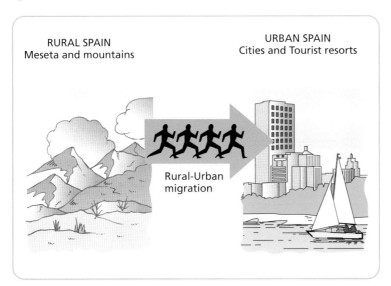

RURAL SPAIN
Meseta and mountains

URBAN SPAIN
Cities and Tourist resorts

Rural-Urban
migration

Figure Q2.4 Migration from rural to urban areas

Look at Figure Q2.4. What effects will this pattern of migration have on rural communities and urban areas?   *(5 marks)*

Intermediate 2 2002 4bii

Answer

There won't be many people in rural areas (✓). There will be many changes from high migration levels to towns and cities from rural areas. There will be many people in cities and areas may be overcrowded (✓). There will be a lot of urban sprawl into the countryside as many more houses will have to be built outside the town or cities (✓) rural communities only have much older people, (✓) the older generation will move out of the town for more quiet lives away from the busy traffic of the towns and cities (✓).

Questions and *Answers* continued ➤

Questions and **Answers** continued

Comments and marks obtained

The answer obtains marks for referring to the depopulation of rural areas, the migration to towns and the impacts of overcrowding and urban sprawl. Further marks are obtained for referring to the population structure namely, 'older people left' and that older people may leave the towns for the quiet of the countryside. The answer has 5 valid points for **full marks**.

Change in Urban Areas

Key Idea 4

You should know about change, problems and policies in urban areas and be able to illustrate your knowledge by referring to an urban area in an EMDC and an urban area in an ELDC.

Key Point 9

You should be able to identify land use zones in a city on an Ordnance Survey map or diagram. You should also be able to describe the main features of different land-use zones found within an Economically More Developed Country city which you have studied.

Figure 2.8 shows a summary of the main zones within cities. You should be able to detect a direct link between land use and land values.

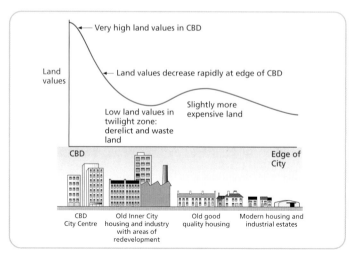

Figure 2.8 Land use zones and land values

Zone 1 The Central Business District

◆ Those functions which provide important services for residents of a settlement are usually located at or near the centre of the city. This area is called the Central Business District (CBD).

◆ You may be asked to identify this area on an OS map by indicating certain map squares. You may also be asked to explain your choice. You would discuss the types of buildings, the pattern of streets (usually narrow or in a grid iron arrangement), the fact that major roads converge their giving greater accessibility and that there are bus and railway stations and public buildings such as town halls.

Zone 2 Industrial zone

◆ Old industrial areas with factories (works) are usually located close to the CBD.

◆ The industrial zone is normally serviced by main roads and railway lines. This provides easy accessibility, allowing factories to bring in their raw materials and distribute their finished products. Easy accessibility also allows workers to get to their workplace easily.

◆ The industrial zone can be identified on an OS map by the shape of the buildings (usually large blocks) and words such as *works* printed beside the buildings.

◆ In older cities, this zone may contain heavy industries or may have gap sites in which old industries, which have since closed, were located.

Zone 3 Low cost housing

◆ Low cost housing is often found in the areas round the industrial zone. The area around factories is usually not the most pleasant to live in, so housing around this area is often cheap and originally may have been built for workers and their families.

◆ On an Ordnance Survey map the streets may form a grid-iron pattern, with narrow streets. The density of housing will usually be high, consisting mostly of tenements or terraced houses. Population density will also be high in these areas and many of the houses may be fairly old.

◆ Zones 2 and 3 make up the area known as the inner city.

◆ In most inner cities, there are areas with derelict buildings and 19th century housing which have since been demolished and turned into brownfield sites.

◆ Some of these sites have been replaced by warehouses, new housing, roads and railways.

Zones 4 and 5 Medium and high cost housing

◆ As people can afford to, they tend to move further away from the town centre. As a result most towns and cities have different quality of residential areas namely low, medium and high cost areas.

◆ In zone 4, streets may have a curvilinear pattern with cul-de sacs and houses may consist of a mixture of tenements, terraced houses and semi-detached houses. House prices will be higher than those in zone 3 and many will be owner occupied.

◆ Zone 5 may have a mixture of detached and semi-detached villas and bungalows and will also be more expensive. The streets may be wider and the density of housing will be much less than in other residential zones.

◆ On the fringes or outskirts of cities and towns there are areas which have been used for newer, more modern industries and out-of-town shopping areas.

◆ Some cities, especially in Scotland, have council house schemes with lower cost housing built in these areas.

Zone 6 Commuter zone and urban fringe

◆ Further away from the edge of the city there may be small villages which during the 1960's became popular for people who preferred to live well away from the city and travel in and out daily to their work.

◆ On an OS map these settlements will be located outside the main built up area and may be linked by an A-class road or motorway. These are known as *commuter settlements*. Their main function is residential. Very few of the residents actually work in the settlement.

Key Point 10

You should be able to discuss the main changes affecting EMDC cities.

The main changes which have occurred recently in EMDC cities include those which have affected housing in terms of inner city renewal, gentrification, ghetto areas and fringe developments.

Inner city renewal

◆ The three main types of housing (low-, medium- and high-cost) have all changed significantly in most settlements in Britain.

◆ Many town and city councils have built areas of council houses both in the inner city and on the outskirts. These have included tenements, terraced houses and high rise flats.

◆ Many residents have had the opportunity to move from older houses, which have since been demolished, to new houses in council house schemes. This process is called *inner city renewal*.

◆ As part of this process city centre areas or central business areas and old industrial zones have also been involved in substantial change. This change has involved relocating industry from inner city areas to new industrial estates on the outskirts of cities. These new estates, however, do not always employ as many people as the old industries and unemployment often remains a problem.

◆ There has also been a shift of population from the city to country areas resulting in the loss of trade within the city.

◆ Changes to transport networks, such as ring roads and one-way road systems, in and around the CBD affects many cities in Britain and Europe.

◆ Some businesses situated on major routes suffer through lack of customers because drivers can no longer park on main roads.

◆ Regeneration schemes in some inner city housing areas have radically changed the quality of housing and the local environment of many housing zones in cities.

◆ Older properties have been demolished and replaced with new housing. In the UK this has included the demolition of high-rise flats which were built during the 1960s. Similar changes took place in cities in France and Germany. Many families and older communities were broken up and redistributed to areas throughout the city.

Gentrification

◆ Gentrification is the process whereby property in both the CBD and inner city which was formerly used for businesses (such as offices, warehouses, and even fire stations) have been converted into modern flats.

◆ Those who buy these properties have reversed the trend of commuting. These are people who prefer to live within the city, enjoying the proximity to their place of work and the various services offered by the CBD.

◆ Property developers have recognised that there is a market for this kind of residence and have renovated older properties to accommodate the demand for houses in these areas.

Ghetto areas

◆ Ghetto areas are areas in cities where large numbers of a particular ethnic group are concentrated.

◆ Reasons for this include the availability of cheap housing and the ability to live in a community with similar customs, culture and language. These are especially important to new immigrants.

Fringe developments

◆ Fringe areas of cities refers to rural zones immediately on the outskirts of cities. These areas attract developments such as housing estates, retail parks and modern industrial estates due to the availability of land for building. Fringe areas have grown considerably.

◆ Large numbers of people were prepared to pay the cost of increased travel from fringe areas in terms of time and money to live in what they considered a better environment.

◆ Small rural villages have been effectively colonised and have become commuter settlements. The populations of these former rural villages have often grown enormously. Residents commute to work and purchase most of their needs outwith the villages.

◆ In other cases, large numbers of people from cities have been offered the chance to move to New Towns in fringe areas with the promise of employment in addition to housing. People living in crowded congested cities often preferred to live in cleaner, less restricted rural areas.

Key Point 11

You should know about land use conflict in rural urban fringe areas and policies and strategies adopted to resolve them.

Conflict in rural and urban fringe areas

◆ Property speculation and compulsory purchase of land by developers has led to a decline in the quality of fringe areas. This has resulted in loss of farmland and recreational land.

HOW TO PASS INTERMEDIATE 2 GEOGRAPHY

- As the commuter belt expanded there has been increased demand for new houses. This in turn has led to huge increases in the volume of traffic on rural roads.

- The loss of population due to development of new towns and commuter belts creates financial problems for city councils and the loss of business to city centre shopping areas.

- Fringe developments such as housing and new industrial estates has led to urban sprawl and loss of land which was previously used for rural purposes.

Policies and strategies adopted to resolve problems

- Green Belt legislation passed in the 1950s in Britain introduced strategies involving planning laws which could restrict developments such as housing, industry, landfill sites and recreational centres.

- Smaller towns and villages were identified for growth. Throughout Britain new towns and overspill areas were used as growth centres to limit further development within the rural urban fringe.

- However, small rural villages remained popular with people wishing to leave the city and were still a target for developers. Limitations on the number and type of buildings have kept the rural environment protected from the worst excesses of city developers.

- The needs of industry and opportunity for offering employment had to be balanced with the desire to protect the rural urban fringe.

Transport Systems

Key Idea 5

Referring to an example from an EMDC city and an ELDC city, you should know about the main changes to transport systems, the problems they create and efforts taken to reduce them.

Key Point 12

You should know about the quality of public transport and commuter zones in EMDCs and the consequences of these developments.

Important aspects which you could refer to might include:

- In Britain, during the 1960s, many small rural villages became popular for people who preferred to live well outside the city and travel on a daily basis to the city for work and shopping.

- These villages are known as *commuter settlements* and the areas in which they are located are termed commuter zones. The main feature of these villages is that their main function is residential. Very few residents actually work in the settlement or immediate area.

- Often the character of these villages changed completely both in form and function. Instead of existing as small rural settlements, many of these villages now have housing estates and their populations have increased enormously.
- The world's largest cities attract millions of workers from outside the city each day. Some of these cities have developed particular types of public transport to cope with these large numbers of people such as the underground train systems of Paris, London and New York.
- During peak hours the road system and public transport systems become heavily congested and accidents or break-downs on major routes can cause chaos.
- In narrow streets with tall buildings in some areas of cities, there is a serious build up of air pollution which is very damaging to the health of city inhabitants.
- Pollution from commuter traffic is a significant contributor to the increase in greenhouse gases in the atmosphere and contributes to global warming.
- Large areas of valuable land have to be used for car parking.
- New roads, which are very costly, have to be built and this often involves demolishing housing and breaking up long established communities.

EMDC cities

In EMDCs the most significant changes to transport systems include:

- new roads, motorways, by-passes and ring roads built around and through cities to cope with the very heavy increases in traffic;
- improvements in the quality of public transport;
- introduction of ring road systems in cities;
- taxing vehicles using inner city area of cities;
- introduction of park and ride schemes for commuters.

Ring road systems

Figure 2.9 shows the orbital road system around London which is based on the M25 motorway.

- This road system provides a convenient way around London and links with the major motorways and Channel ports. This has eased traffic congestion within the city.
- There are similar ring road systems operating throughout many parts of the Developed world. All of these are aimed at reducing the volume of traffic travelling in and out of major city areas.
- Traffic using these systems can bypass city areas thus reducing problems of environmental pollution, road maintenance, traffic congestion and road accidents in inner city areas.

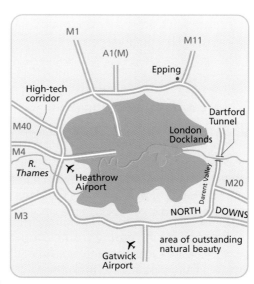

Figure 2.9 The orbital road system around London

65

However, ring roads have created problems for the surrounding areas including:

◆ increases in property values near ring roads (because industry and commuters see the advantages of living close to motorways);

◆ development pressures on land in the Green Belt and the destruction of many attractive areas of countryside such as the North Downs of England;

◆ industry and commerce being encouraged to invest more in economically more attractive areas such as the South East of England rather than as opposed more economically depressed areas;

◆ local wildlife habitats have often been destroyed to make way for these new roads;

◆ increased levels of local pollution levels on ring roads due to traffic fumes.

Road pricing schemes

◆ In some European countries road users are charged a toll on major transport routes, especially motorways.

◆ These charges are directly related to the distance travelled. There are different rates for different types of vehicles such as private cars, road haulage vehicles and tourist buses. These charges contribute to the cost of building and maintaining the motorways.

◆ Apart from toll charges on bridges such as the Forth Road Bridge, the Severn Road Bridge and tunnels such as the Dartford Tunnel in London, there are generally no charges, as yet, for those using Britain's motorways. However, drivers are charged for using a new section of the M6 near Birmingham.

◆ There are however *congestion charges* for those driving on inner city roads in London and there are government and local authority plans to introduce similar charges in other British cities.

◆ Congestion charges are part of a government strategy to reduce the volume of road traffic to reduce congestion and improve air quality.

◆ The charges operate through the use of highly technological computer and camera systems which record and log vehicles passing through delimited zones. Vehicle owners are sent a bill and if road charges are not paid within a specific time limit, fines are issued on top of the charges.

◆ Revenue raised from congestion charges is used for improved public transport and for road building.

Increasing traffic congestion

Traffic congestion is one of the biggest transport problems found in most cities in the world, especially in EMDCs. The volume of traffic has increased tremendously with huge amounts of private and commercial vehicles entering and leaving city areas through the day. The problem is at its greatest during the morning and evening rush hours. However, weekend traffic is often just as bad. This has led to problems of congestion, accidents, damage to roads, high costs of road maintenance and disruption to the life and work of the city.

Reasons for increasing traffic congestion in EMDC cities

◆ Increased volume of motor vehicles on roads linked to commuter traffic from rural and urban fringe areas at peak times.

◆ Poor quality roads within inner city areas unable to cope with volume of traffic. Older cities often have road patterns designed and built before mass transport was an issue.

- Lack of adequate public transport services in most cities.
- Parking on main roads and lack of parking facilities within city centre.

> ## Key Point 13
>
> You should know the main measures adopted in cities to reduce traffic congestion.

Measures to reduce traffic congestion
- Changes to road systems such as ring roads; one-way systems; use of bus only lanes; contraflow systems; widening of roads; building by-passes.
- Parking restrictions prevent parked cars blocking main routes (enforced by police and traffic wardens).
- The provision of alternative means of transport, to encourage drivers to leave their cars outside the city and travel into cities with linked light railway or underground rail systems.
- Park and ride schemes whereby car users park their cars outside the city and use public transport to complete their journey into the city centre.
- Cheap public transport on buses.
- Congestion charging and increased parking charges.

Problems created by these measures
- Preventing people from coming into the city can create problems for retailers which rely on customers for their trade.
- Offering alternatives such as out-of-town shopping centres may help solve traffic problems, but providing these centres often leads to the decline of the Central Business District from which a large proportion of local government finance is obtained.
- Authorities have to balance their solutions very carefully. National government policies recognise this.
- New proposals in Britain include imposing tolls on cars entering the city, increasing fuel prices and upgrading railways and other public transport services.

Questions *and* Answers (?)

Question 2.5

Figure Q2.5

'Traffic congestion is a major problem in cities in Economically More Developed Countries (EMDC)'.

Look at Figure Q2E and the statement above. Referring to a city you have studied in an EMDC, describe in detail measures which can be taken to reduce the problem of traffic congestion.   *(6 marks)*

Intermediate 2 2005 2f

Answer

In Edinburgh which is in an EMDC measures have been introduced to reduce the problem of traffic congestion such as bringing in park and ride systems (✓) whereby people park their cars outside the city centre and travel in on a bus for a low charge(✓). It costs less than it would to park a vehicle in the city centre and it means one bus carrying many people on the road instead of many cars carrying one person cutting down on congestion(✓).

In London a tariff has been introduced which is a charge people will have to pay if they want to drive their cars into the city centre (✓).

More bus services and bus tours have been set up to encourage people to use them instead of driving in and congesting the city centre.

In Perth a large part of it has become pedestrianised to stop cars driving right through large parts of the CBD. Instead they can only drive around it and on certain roads through it therefore reducing traffic congestion (✓).

Questions *and* Answers *continued* ➤

Questions *and* *Answers* *continued*

Comments and marks obtained

Although only asked for one city studied, this answer refers to three which is acceptable. Marks are obtained for reference to park and ride schemes and explaining how they operate. A third mark is gained for pointing out the advantage of buses over cars. Mentioning the tariff charges in London gains a fourth mark. The next statement on buses is too general and basically repeats the comment on park and ride made earlier to gain any more marks.

The final comment on pedestrianised zone in Perth earns a final mark giving a total of **5 marks out of 6** for the answer.

Transport problems in cities in ELDCs

ELDCs often have similar problems to those in EMDCs caused largely by mass movements of people during the working day. However, the problems are often made even worse due to lack of development.

◆ Insufficient money to pay for the ever-increasing costs of providing a transport infrastructure which works.

◆ A transport network system which is inherited from previous colonial rulers.

◆ An emphasis on building large ports, high grade roads and railways used for the transport of heavy and bulky natural resources, instead of a transport system designed for mass movements of people. Developments such as these have left little money to improve access to and from the poorest parts of these countries. ELDC cities need efficient transport networks to aid their development.

◆ Present networks in Africa and South America encourage a polarization of resources. This means that more and more of the wealth of these countries is concentrated in a very small number of areas, usually in the large cities. This trend has resulted in many rural areas in these countries being deprived, resulting in high levels of unemployment and poverty.

In North Africa, attempts have been made to develop a new trans-continental highway system. This has been done in an effort to improve communications and to help the countries of North Africa develop trade links. Improved trade and communications will eventually help in the overall development of the economies of these African countries. Figure 2.10 shows the proposed North West African Highway and some of the other major routes.

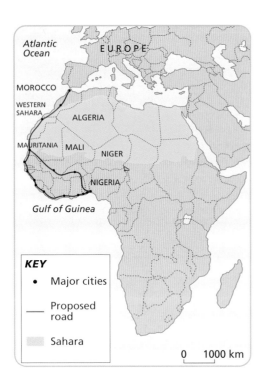

Figure 2.10 The proposed North West African Highway

Strategies to reduce transport problems in ELDC cities

- ◆ Increase provision of public transport including buses and trains.

- ◆ Large scale use of bicycles. This is popular in many countries such as India, South East Asia, and in South America.

- ◆ Construction of new roads to produce a more efficient transport infrastructure in many developing countries. These efforts have been hindered by a lack of sufficient funds.

CBDs and Inner Cities

Key Idea 6

You should know about changes to retail services in cities. With reference to an EMDC city you have studied you should be able to describe and suggest reasons for change within the Central Business District.

Main changes in EMDC central business districts

- ◆ Central business areas are affected by high rental costs and declining custom. This can lead to the closure of some shops. Only the most successful retail outlets can afford to stay in the CBD.

- ◆ Large retail parks and trading estates have been built on cheap land on the outskirts of major cities. These estates are well connected by a good road system. They also offer the advantage of free car parking.

- ◆ Cities have responded by building large covered shopping centres with car park facilities within their central business areas to attract customers back to the city. These efforts have been successful in cities such as Glasgow, Manchester and Paris. Improvements and changes to roads and streets can ease problems of traffic congestion.

◆ New office blocks replace older buildings such as bus and railway terminals. Former industrial and dockland areas are radically altered and are replaced with new conference centres, housing, business parks, museums and leisure complexes.

Key Point 14

You should be able to describe the policies adopted to resolve the problems related to the CBD and inner city areas.

Policies to resolve problems in the CBD and inner city areas

◆ Repopulation of city centres. This is achieved by converting old buildings such as warehouses and former public buildings into flats. These flats are attractive to people who want to reduce the cost of commuting every day and have quick access to the facilities of the city centre.

◆ Demolition of abandoned industrial sites and the rehabilitation of older housing areas. This helps improve the inner city areas and make them more attractive.

Environmental quality in cities

Key Point 15

You should be able to describe and suggest reasons for changes related to environmental quality in cities.

Dereliction of older buildings

◆ Old dilapidated factory sites have been demolished in many cities and have been replaced by new trading estates. Once the buildings were demolished, the areas left were known as brownfield sites.

◆ The environments of these areas have been transformed and visual pollution has been greatly reduced as a result. Modern buildings with landscaped tree lined avenues have replaced the ugly, unpleasant factories and warehouses.

◆ The estates are accessed by new road systems and have car parking for workers and customers.

Urban decay within cities

◆ Many parts of cities have buildings which have fallen into disrepair and decay through age and lack of care. These buildings may be industrial buildings, housing, office blocks or even shopping areas. Their visual appearance may be unsightly and more importantly they may be unsafe for use or habitation. Often these buildings are left in a poor state for many years before action is taken.

◆ Common sites of decay include inner city housing areas and former industrial sites near railway lines, canals, along the banks of rivers and at the edges of the central business area.

- Inner city areas also suffer from traffic congestion, narrow streets, poor quality housing, pollution and declining industrial areas with derelict sites causing urban blight.
- Urban decay is often made worse because of a lack of money for demolition and redevelopment.

Key Point 16

You should also know about problems caused by environmental changes within inner cities and the policies and strategies adopted to solve them. The problems include dereliction, urban decay and pollution.

Inner city renewal schemes

- Old factories and housing are replaced by new offices, industrial parks and retail parks.
- New buildings include office blocks, new housing, recreational centres and shopping centres.
- New road systems are constructed to make these renewal areas more accessible.

These changes have often been achieved through the financial assistance of government grants and European Union grants.

Pollution within cities

- Pollution includes urban blight caused by derelict buildings, litter, vandalism, air pollution from industries and traffic and river pollution from industries discharging waste.
- Rivers such as the Clyde and the Thames which flow through Glasgow and London respectively suffered for many years from pollution caused by the dumping of industrial waste from factories along the banks of the rivers. In many rivers this problem became so bad that hardly any fish life survived.
- Spillages from oil terminals and waste from major industries such as the famous Clyde shipyards also badly affected the quality of the water of the rivers.
- Pollution from sewage disposal units, motor vehicles or through neglect can accumulate to such an extent that it can not only destroy the appearance of some parts of cities, but may also create a health hazard.

Policies adopted to solve the problems of damaging pollution

- Introduction of smokeless zones, building of high chimneys and the banning of industries which are the source of obnoxious smells or pollutants.
- Implementation of laws and regulations to monitor and control pollution. These laws are enforced by environmental health departments employed by most city councils. Although this is costly, the benefits are obvious to the residents of cities.
- City authorities must continue to employ these measures every day to ensure high standards of environmental quality are maintained.
- All of these problems have been successfully tackled in cities throughout the developed world by measures outlined above.

Key Point 17

Referring to an example from an EMDC and an ELDC city, you should be able to describe and explain differences in the provision of safe water, sanitation and waste disposal particularly between urban and rural areas.

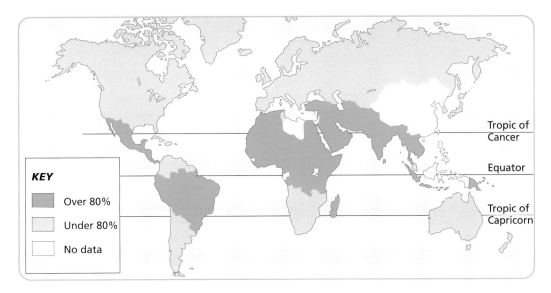

KEY

- Over 80%
- Under 80%
- No data

Figure 2.11a Percentage of population with access to safe water

Figure 2.11b Safe water and sanitation in ELDCs

Sanitation within cities

◆ As Figure 2.11 shows, access to safe water is a major problem in ELDCs. Lack of clean, safe water leads to the contraction of many water borne diseases such as typhoid, schistosomiasis, cholera and diahorrea.

◆ Lack of clean water in these countries may be due to physical factors such as drought or economic factors such as lack of funding for sanitation and water supply to households.

◆ In ELDCs water supply and sanitation problems tend to occur more in poorer agricultural areas which rely on local rivers and streams for water. These sources often become contaminated and breeding grounds for disease carrying insects such as mosquitoes and blackflies.

◆ Urban areas in ELDCs may have better sanitation and water supplies than rural areas although they may not be available to all residents, particularly those in squatter areas or shanty towns.

◆ Millions of tons of residential and industrial waste must be disposed of by city authorities. Some of this waste may be incinerated in furnaces whilst much of it is buried in landfill sites. This can create major problems, especially if chemicals in the waste seeps into the underlying water table.

◆ Waste disposal is a major problem in both EMDC and ELDC cities. Millions of tonnes of waste affect city water supplies and ultimately the health of city inhabitants. The problems are greater in developing countries due to the lack of funds to properly finance waste disposal.

◆ In some cities the city area has been extended near the coast by landfill waste sites which form the foundations for new buildings.

◆ Some countries try to dispose of their waste by dumping millions of tons of waste in surrounding seas and oceans. This leads to massive pollution of these sea areas. This kind of action is vigorously opposed by organisations such as Greenpeace who demonstrate against authorities dumping waste in seas and oceans.

◆ Waste disposal through incineration can increase levels of air pollution.

Case study
EMDC city – Inner city change in Paris

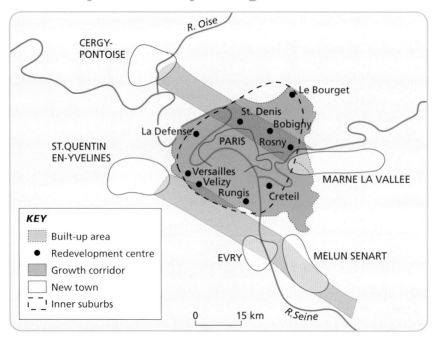

Figure 2.12 Change in Paris

- In Paris several major changes have taken place within the inner city. Most of these involved changes to housing. Old tenements which often lacked basic facilities were demolished and replaced by new and renovated housing.

- The local authorities in Paris set up regeneration schemes such as La Defense. These schemes involved the demolition of dilapidated tenements and replacing them with residential buildings in tower blocks up to 40 storeys high. 35 000 people currently live in these high rise blocks.

- La Defense also provides new offices and shopping facilities in addition to the improved housing. The area is a traffic free zone with roads and car parks located underground and the area is connected to the city centre by the underground metro rail service.

Gentrification of the inner city area of Paris

- In Paris the worst slum areas were pulled down or renovated and replaced by improved housing. Wealthy middle class people, working in central Paris, bought some of this housing cheaply.

- Houses were improved by the new owners and this encouraged other to improve their property. Gradually some parts of the inner city in Paris became fashionable to live in for the more wealthy, including areas such as Bellville and Bercy.

- Advantages of living in these areas included easy and cheap access to work, shopping and entertainment in the central area. Commuting costs were reduced making it easier to afford gentrified property in the inner city.

Ghetto areas in Paris

- As redevelopment of the inner city took place in Paris, other areas, such as Sarcelles, housed large numbers of people in large housing schemes. These schemes became unpopular due to the lack of jobs, shops and public transport.

- As people moved away from the area they were replaced by immigrant families, especially from North Africa. Many immigrants found it difficult to find accommodation in any areas other than the poorest.

- Since many immigrants preferred to live within their own ethnic community, with people of the same language, culture and customs, the percentage of the immigrant community increased greatly. Some of the housing schemes became ghetto areas.

- The lack of proper housing forced some immigrants to build shacks and turned areas into shanty towns. These are known as *bidonvilles* in France.

Fringe developments near Paris

- Five new towns were built within a radius of 25 kilometres of Paris to address the serious housing problems in the city.

- Up to 200 000 people currently live in new towns such as Cergy-Pontoise.

- New housing and industry was encouraged to set up in these towns and in growth centres such as Roissy, outside the city.

Regeneration in Paris

- As part of the regeneration of the inner city of Paris, the largest shopping centre in Europe was opened in La Defense in 1981. In recent years this centre has become less popular with shoppers due to traffic problems.

◆ Many shoppers now prefer to shop in nearby suburban shopping centres or in the central area of Paris.

◆ A number of large hypermarkets have been built along main roads at the edge of Paris.

◆ Whilst these shops were initially popular due to the wide range of goods sold and the competitive prices, city authorities are now less keen to build more of them since they contribute to urban sprawl.

Case study
EMDC city – Commuting in New York

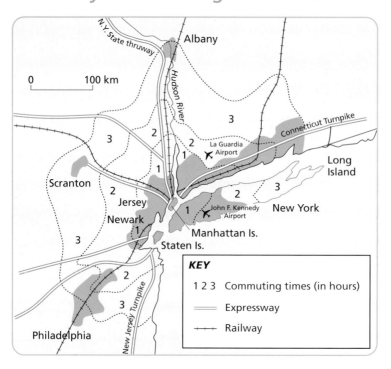

Figure 2.13 Manhattan's commuter hinterland

◆ With a population of over 16 million people and a daily total of 4 million commuters, New York has to meet the challenge of commuter traffic on a massive scale.

◆ Two million commuters per day have to get into Manhattan Island. 75% of commuters in New York use public transport, but that still leaves a huge number of cars trying to get to the island every day. Since there are only a few tunnels and bridges linking the island to the mainland, this becomes a major problem for the city.

◆ New York has built expressways above built-up areas to cope with the traffic. This has cost billions of dollars and the city gains no revenue from taxing commuters.

◆ These problems are typical of those faced by major cities throughout the developed world.

Questions and Answers

Question 2.6

For any city you have studied in mainland Europe, describe the steps taken to solve typical inner city problems.   *(5 marks)*

Intermediate 2 2004 4c

> **Answer**
>
> *Paris, France, is an area I have studied with typical inner city problems. To solve these problems buildings that are old and slum tenements have been knocked down (✓) and rebuilt in high rise flats (✓). Shops and offices etc., have been moved to the outskirts. More shops or hypermarkets have been built in the outskirts (✓). This is to stop car pollution and noise in the inner city area (✓). There is also park and ride schemes to save as many vehicles travelling through Paris(✓).*

Comments and marks obtained

By using Paris as a real example the candidate has ensured that the answer can gain full marks. The references to the demolition of slum areas, replacing them with high rise flats, shops and hypermarkets being built on the outskirts gains three marks. The final comments on efforts to prevent noise pollution and park and ride schemes to reduce traffic in Paris also merit a further two marks, making a total of **5 marks out of 5** for the answer.

Key Point 18

In Economically Less Developed Countries, changes to urban areas have involved developments related to high security areas, squatter areas and shanty towns.

Key Point 19

For an ELDC city you have studied, you should be able to describe and explain various social, economic and environmental problems resulting from the city's growth and efforts made to resolve them.

HUMAN ENVIRONMENTS

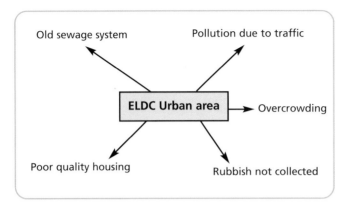

Figure 2.14 Urban problems in Economically Less Developed Countries

Social problems in ELDC cities

◆ High expectations of incoming population to cities such as Sao Paulo and Calcutta are not always realised.

◆ Lack of housing often results in homelessness. Families consequently have to live either on the streets, in squatter areas or in shanty towns. Many houses lack basic amenities, such as electricity and particularly a supply of clean water.

◆ Lack of proper facilities such as clean water and electricity and sewage disposal schemes lead to health problems, such as the spread of diseases such as typhoid and cholera. In the worst cases people use polluted wells and open pipes.

◆ Conditions generally may be very unhygienic and unhealthy. Infant mortality rates often increase and life expectancy decreases.

Economic problems

◆ Lack of industry affects employment opportunities. This creates problems of low income and poor standards of living involving a lack of food, poor education and health care. Housing quality is poor due to lack of investment.

◆ Crime rates are often very high due to extreme poverty. Local police are often unable to deal with the number of crimes committed due to lack of manpower.

◆ Middle class residents outside the poorer quarters live in high security areas which have secure entrances and are patrolled by security guards. Access is restricted. The cost of providing this level of protection is very high and cannot be afforded by the majority of the city's population.

Policies adopted to solve problems of ELDC city growth

◆ Funding is obtained by city authorities from sources such as United Nations agencies such as United Nations Education Social and Cultural Organisation (UNESCO), World Health Organisation (WHO) and World Bank to provide money for medical and educational services for local communities.

◆ City authorities may introduce primary health care schemes and local self-help schemes, particularly in the shanty towns. In some cities efforts have been made to rehouse people in new town areas.

◆ Improvements to water supplies in many residential areas have been treated as a priority.

◆ Self-help schemes are promoted to improve local infrastructure, using local labour.

◆ Materials have been supplied to help improve the basic fabric of the poorest quality houses.

◆ Financial incentives have been offered to residents to move to other parts of the city. (However, this has often been due to the expansion of commercial enterprises and the need for land rather than efforts to help residents of shanty towns.)

Rural Change

Key Idea 7

You should know about change, problems and policies in rural areas and be able to illustrate your knowledge by referring to a rural area in an EMDC and a rural area in an ELDC.

Key Point 20

You should know about agricultural systems in an EMDC and an ELDC and the type of landscape associated with these systems. You should also know how these systems and landscapes can change, the reasons for the changes, and the implications of these changes.

Agricultural systems involve inputs, processes and outputs.

In farming, **inputs** include:

◆ **physical** factors such as climate, landscape and soils;

◆ **human** factors such as labour, land ownership systems and cultural background;

◆ **economic** factors such as capital investment, technology, use of seeds and fertilisers, government influence, transport and markets.

Depending on the type of farming, various **processes** occur between the input and output stages. These include:

◆ preparing the land for crop growing, ploughing, seeding, harvesting;

◆ tending livestock;

◆ transporting produce to market;

◆ maintenance of equipment;

◆ use of artificial drainage and irrigation;

◆ use of labour and machinery on the farm.

Outputs include arable crops, livestock, farm produce such as dairy products and any profit made from selling produce.

Key Point 21

Referring to a rural area in both an EMDC and an ELDC you have studied, you should be able to describe agricultural change and comment on the benefits and problems caused by these changes.

Change in rural areas in Economically More Developed Countries

As with other economic enterprises, farming has changed greatly over the last 50 years. These changes have affected methods, organization, farm output, labour, farming landscapes and the overall status of farming within the economy. Several factors have been important in bringing about these changes including diversification, urban sprawl and tourism.

Diversification

◆ In order to reduce costs and increase the profitability of farms, farmers have increasingly added new non-farming land uses to their farms.

◆ Much of this is linked to the leisure and tourist industry. Areas of farms are now used as golf courses, mountain bike trails and 4×4 rally tracks.

◆ Cottages once occupied by farm workers, made redundant, have been converted to holiday homes for tourists. Areas have been set aside for camping and caravan sites. Some farm houses offer bed and breakfast accommodation. All of this makes the farmer less dependent on just their farming income.

◆ Problems which result from this approach include the fact that smaller farmers find it more and more difficult to compete with larger enterprises. Without increasing capital input it is hard to match the efficiency of the bigger units. Some smaller farmers have therefore had to sell their farms to their larger competitors.

◆ Some farms have changed the types of crops they grow, from traditional arable crops to higher value crops such as strawberries and raspberries, and peppers and tomatoes. These changes often require new farming techniques such as the use of polytunnels.

◆ Those formerly employed in farming have lost their jobs due to increased mechanisation and have had to find alternative employment or leave the area altogether in search of work. Population density has therefore decreased in some areas.

◆ In Europe the European Union (EU) has had a major impact on agriculture. A variety of policies including quotas, set-aside policies, subsidies and other legislation has led to many developments, some for the better, some for the worse.

◆ Output has increased greatly but this has in the past led to huge surpluses of produce resulting in the infamous butter mountains, apple mountains and wine lakes. Produce is often stored in barns in order to reduce supply and maintain prices.

◆ Farmers have received payments from the EU for setting aside land rather than growing crops. The cost of this has been met from taxpayers in member countries of the EU.

◆ The EU has also paid development grants to poorer agricultural areas such as Southern Italy to help farmers to modernise their farms.

Urban sprawl

♦ Property speculation and the purchase of land by developers has led to a loss of farmland and recreational land. This has resulted in a decline in the quality of the rural-urban fringe around many cities. This has caused conflict between traditional rural activities such as farming and new developments.

♦ As the commuter belt expands there is ever-increasing demand for new houses. This in turn leads to increase in traffic on rural roads.

♦ Green Belt legislation introduced in the 1950's was an attempt to curb development in rural fringe areas and has been quite successful. Green Belt strategies involve planning restrictions and restrictions on developments such as housing, industry, landfill sites and recreational centres.

♦ Smaller towns and villages were identified for growth. For example, in Central Scotland new towns and overspill areas such as East Kilbride were used as growth centres to limit further development within the rural urban fringe.

♦ However, small rural villages remain popular with people wishing to leave the city and are a target for building developers.

♦ A balance needs to be struck between the desire to protect the rural-urban fringe and the need for employment for local populations.

♦ Limitations on the number and type of buildings have kept the rural environment protected from the worst excesses of city developers.

Impacts of tourism

Tourism in rural areas in EMDCs has a number of conflicting impacts.

♦ Many urban dwellers make frequent day or weekend visits to the countryside, often by car. For example, car-owning Londoners may go to the North Downs, whilst people from Sheffield may travel to the Peak District.

♦ Visitors now engage in different and more diverse activities than they did in the past. Cycling and walking have replaced 'going for a drive' and picnics.

♦ Access to the countryside has improved with initiatives such as National Trails and the Right to Roam Campaign. This has led to an increase in bed and breakfast accommodation, tearooms, pubs and other services.

♦ Mountain biking, off-road driving, orienteering, hang–gliding and water based sports have grown in popularity.

♦ Other changes in tourism have included the growth of theme parks, holiday villages and farm tourism. Farmers have converted barns and cottages to holiday lets and use meadows for caravan and camping sites.

♦ Farm tourism is especially well developed in Europe (particularly in Italy and Austria), and in Wyoming in the USA dude ranches are very popular.

♦ In places such as the Lake District and upland Wales, where incomes have declined from farming, farm tourism has been particularly beneficial to farmers.

♦ Commercial attractions such as steam railways, wildlife parks and rural museums have also developed.

♦ Second home ownership and timeshare complexes have also developed widely in rural areas. However, these developments have created conflict in many rural settlements.

Problems resulting from tourism

◆ As well as the benefits, problems have accrued from rural tourism such as competition for housing, traffic congestion, conflict with farmers, overuse of local water supply, litter increase and damage to the physical environment through erosion of hillsides and footpaths.

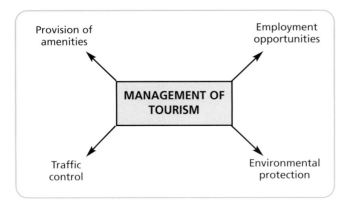

Figure 2.15 Management of tourism

Measures and policies adopted to tackle problems caused by tourism in rural areas

◆ Public bodies such as National Park Authorities make efforts to ensure that tourism is sustainable and the damage is minimal. Legislation to restrict access, the use of zoning, fines and limits on certain developments such as inappropriate industrial and residential land uses are employed.

◆ In Dartmoor National Park areas of high and low tolerance have been identified as a means of releasing pressure on the area through tourism.

◆ Country Parks were established in 1968 and now act as a focal point for countryside visits. Other honeypot sites have also been developed to attract large numbers away from more sensitive areas.

◆ Another measure used to manage visitor pressure is that of dispersal of tourists to spread their effect more evenly. This includes designing scenic drives which take tourists to less frequented areas.

◆ Education centres have also been set up in Parks to educate people on how best to use and protect the countryside and its vegetation and wildlife.

◆ These measures have been quite successful in reducing the problems created by increasing use of rural areas by tourists.

Examination questions

Typical questions on these topics would include resources such as graphs showing trends in workforce, size of fields, output and number of farms. You might then be asked to describe the trends and then to explain these trends. Other questions may show two farming landscapes at different periods in time, and ask you to explain the differences or give reasons for the changes which have occurred.

◆ If you simply describe changes, when asked to explain, you will forfeit marks.

- ◆ You can mention the main changes but you must give reasonable explanations as to why these changes or differences have occurred.
- ◆ If you base your answer on increased mechanization and diversification and develop these points you should gain most of the marks available.

Questions and Answers

Question 2.7

The traditional character of settlement is changing in rural areas of mainland Europe. Explain why this is happening. *(4 marks)*

Intermediate 2 2004 4d

Answer

This is happening because people are buying houses in rural areas as they are so cheap to buy compared with houses in urban areas (✔). They are using them for holiday homes or renting them out to people for a holiday house (✔). Locals are complaining because the houses are probably sitting empty for months and then in summer very busy (✔). Locals will also be annoyed that there will less money for the services in the area if there is nobody staying in these houses throughout the year. (✔)

Comments and marks obtained

The answer successfully discusses the changes, the reasons for them and the effects. References to people buying houses and using them for holiday homes gains two marks. The reference to the effects – houses lying empty and less money for services gains a further two marks giving a total of **4 marks out of 4**.

Key Point 22

Referring to an example of a rural area in an EMDC which you have studied, you should be able to discuss the main changes which have occurred, the reasons for them and the impact which the changes have had on people and the landscape.

Figures 2.16 and 2.17 shows the ways in which farms have changed in the UK between the years 1950 and 2000.

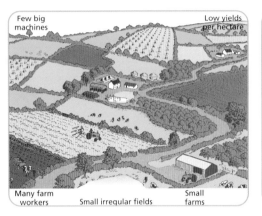
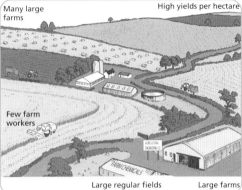

Figure 2.16 Changing farmscapes in the UK

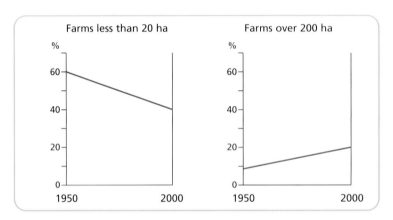

Figure 2.17 Changes in farm size

Changes in farming methods

◆ Increased and improved use of mechanisation. Farms in EMDCs rely heavily on the use of combine harvesters, mechanised milking machines, tractors and other sophisticated pieces of farm machinery. Making use of more machines means that less labour is needed. Buildings have to be built to store the machinery.

◆ Field sizes are increased to allow large machines to operate. This has resulted in hedges being removed and the size of farms increasing. Many farms have amalgamated to increase in size.

◆ Many farmers belong to cooperatives. The cooperatives and individual farmers often hire contracting companies which supply labour and machines such as combine harvesters at harvest time. Hiring machinery and labour is more cost effective for farmers than buying their own machines and carrying out the harvesting jobs themselves.

◆ These contracting companies providing labour and machines migrate northwards during the summer months according to the time of year when the cereals ripen.

♦ Increased mechanisation has resulted in reductions in labour input. This has led to a reduction in population within areas with this type of farming.

♦ Farming methods have become increasingly more scientific with computers, advances in medical care for animals, new improved seeds, chemical fertilizers and insecticides being used on a much wider scale.

♦ Farmers are trained in agricultural colleges to use modern technology and methods to improve output, reduce costs and increase profits.

♦ Different types of seeds (such as disease resistant seeds and seeds with faster growing properties) have been introduced. This has resulted in higher crop yields each year. Yields have also been increased through the use of increased amounts of fertilisers and pesticides.

♦ The recent innovation of genetically modified crops has also been a further change in the farming process, although this step is highly controversial.

♦ Infrastructure has to be continually improved to speed the process of transporting crops to markets. Measures to improve road and railway services are constantly being reviewed with a view to increasing efficiency. For example, wheat is stored in huge elevators which are linked to the railway transport system.

Questions *and* Answers

Question 2.8

Look at Figure Q2.4 earlier in this unit (p 59). Outline the reason for population movement from rural areas to the cities in Spain or any other European country you have studied.   *(4 marks)*

Intermediate 2 2002 4bii

Answer

In France for example La Defence in the Paris basin, people had migrated to cities for more employment (✓) as in rural farming areas jobs would be taken over by machinery (✓). So the farmers would have the work to do only by himself or generations of his family. Whereas younger generations would go to cities for employment and entertainment as there would be more nightlife in cities such as cinemas, clubs and shops (✓). The cities are more central for travelling to different jobs in other nearby places, easier to catch the train or bus whereas in rural areas there isn't as much public transport (✓).

Comments and marks obtained

The answer gives four good explanations as to why people are leaving the countryside, namely employment opportunity in cities, loss of jobs due to mechanisation, entertainment and nightlife available and provision of good transport network. These are sufficient to merit 4 marks out of 4 for the answer.

Key Point 23

You should know about change in rural areas in Economically Less Developed Countries and the impact of new technology and political policies.

There have been various attempts by governments of countries such as India to introduce modern methods and changes to areas which practise intensive peasant farming such as the Punjab in north-east India. These measures have been described as the *Green Revolution*. The main features of the Green Revolution are summarised in Figure 2.18.

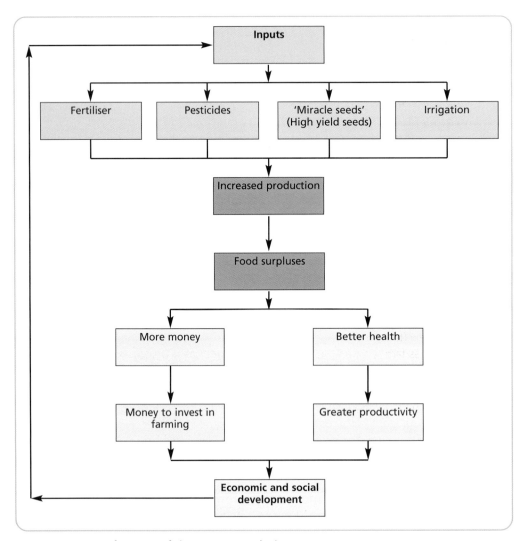

Figure 2.18 Main features of the Green Revolution

In the 1970s and 80s, the Indian government introduced various schemes to improve farming during a programme of 5 and 7 year plans. This involved a range of measures including:

◆ land reform schemes whereby small farms resulting from the land inheritance system have been amalgamated into larger farms;

◆ encouraging farmers to borrow money to improve their farms;

◆ introduction of 'miracle seeds' in order to increase yields;

◆ using chemical fertilisers to improve soil fertility;

◆ increasing mechanisation by using tractors and other farm machinery;

◆ employing agricultural advisers and setting up various training schemes for farmers;

◆ spraying insecticides onto crops to prevent crops being eaten and destroyed by insects;

◆ introducing modern irrigation methods to replace inefficient methods such as inundation canals;

◆ raising the level of technology used on farms such as introducing motorised ploughs;

◆ introducing legislation designed to increase the size of fields and allow the system to use large machinery and become more efficient.

Key Point 24

You should be able to discuss the implications of these changes on people and the landscape in an ELDC and comment on their success or otherwise.

Impact of change on people and landscapes in ELDCs

◆ All of the measures employed in the Green Revolution were designed to increase yield and output from the farms. As a result agricultural production in India doubled during the 1970s and 1980s.

◆ Farmers were encouraged to spend money on fertilisers, pesticides and other new techniques. This spending required government loans and grants, as well as bank loans. However, as farm yields improved, the increased supply meant that the prices received by the farmers for their produce decreased, so they had less money to repay the loans.

◆ Many farmers did not earn enough money to achieve a reasonable income and be able to repay the money they had borrowed. Only the richer farmers benefited.

◆ Increased use of mechanisation left many farmers unemployed. Consequently many farmers had to give up their farms leaving to migrate to other areas or large cities.

◆ Land reform helped to speed up the process of mechanisation and modernisation but deprived many farmers of the opportunity to own their own farm and provide food for their families.

◆ The Green Revolution has been successful in some areas where land reform was successful. In other areas the success rate has been much less and in some cases many farmers have become poorer as a direct result.

HUMAN ENVIRONMENTS

HOW TO PASS INTERMEDIATE 2 GEOGRAPHY

> *Key Point 25*
>
> You should know about change involving migration patterns in rural areas of EMDCs.

Rural push and urban pull factors

The factors which encourage people to leave the rural areas are rural push factors and those which attract people to the towns and cities are urban pull factors.

Rural push factors
◆ Lack of good farming conditions such as poor soils, dry climate, difficult terrain.
◆ Loss of farmland through land reform measures or inability to repay debts.
◆ Lack of employment through increased use of mechanisation.
◆ Low standards of living, resulting from low wages from agricultural employment, poor living conditions, lack of educational opportunities and lack of facilities such as entertainment and other attractions, especially for younger generations.
◆ Quality of health care may be severely limited through lack of hospitals and medical staff. In some cases problems are made worse as a result of natural disasters such as floods, famine and droughts.

Urban pull factors
◆ Availability of employment in a variety of jobs including manufacturing and service industries.
◆ Higher standards of living with higher wages, better housing conditions, better health care, better educational opportunities for younger people and a wider range of attractions such as shops, entertainment and other services.
◆ Possibility of family and friends already living in urban areas offering encouragement.

Case Study
Change in a rural area in an ELDC – intensive peasant farming in India

In India the main system of farming is best described as *intensive peasant farming*. Within intensive peasant farming in India, several different types of crops may be grown depending on local climatic conditions. In wetter areas rice is the main crop. There are several different varieties of rice which can be grown but the most important is wet rice.

The main features include:
◆ fields which are usually very small due to the land tenure system;
◆ rice being planted under water;

Figure 2.19 A peasant farming landscape in SE Asia

- ◆ a lack of mechanisation and therefore a high number of workers working in the fields planting and harvesting crops by hand;
- ◆ the use of irrigation ditches to transfer water to fields, especially in drier areas;
- ◆ water-retaining embankments around fields;
- ◆ terraced hill slopes to maximise the use of land and to conserve soil and water content;
- ◆ the use of animals such as oxen to draw carts and transport crops.

Main changes in peasant farming

- ◆ There have been changes to the land tenure system whereby small farms have been amalgamated into much larger units. Field sizes have increased.
- ◆ Many farmers have sold their land under government schemes.
- ◆ More machinery is used instead of manual labour. More farm workers are unemployed as a result of increased mechanisation.
- ◆ Through the Green Revolution programme, agricultural advisers have provided advice to farmers on new methods of growing crops using irrigation, new miracle seeds and fertilisers.
- ◆ Farmers were offered government grants and bank loans to purchase new materials. Yields have increased but prices for crops have decreased due to supply outweighing demand. As a result many farmers have been unable to repay loans and have lost their farms. Only richer farmers survived.
- ◆ This has resulted in farmers leaving the rural areas and migrating to cities to find work. Populations of rural areas have decreased.
- ◆ Change has brought both benefits and problems to rural areas.

Questions and Answers

Question 2.9

The Green Revolution was an example of change in rural areas in India. For either the Punjab or any other rural area in an ELDC which you have studied:

(i) describe the changes which have taken place. *(4 marks)*

(ii) comment on the successes and failures of these changes. *(4 marks)*

Answer

(i) *In India the Green Revolution has had various degrees of success.*

Land which was previously 'fragmented' has been joined together (✔) and landless workers have been given sections of land (✔). The land owner's land is no longer fragmented but is all in one area (✔).

Machinery has been introduced (✔) and is used in areas, new irrigation methods (✔). Fertilisers and pesticides have been introduced (✔) and their uses have been shown to be successful (✔).

Questions and Answers continued ➤

Questions and **Answers** continued

Answer continued

Farmers have been educated (✓) in the use of these new materials.

New high yielding varieties (HYV) (✓) have been introduced to replace the old seeds.

Comments and marks obtained

The answer contains some comments which are not awarded any marks such as the first sentence. Marks are achieved for the statements on joining of previously fragmented land, workers given sections of land, machinery being introduced, new irrigation methods, fertilisers and pesticides, education of farmers and the introduction of new, high yielding varieties of seed. The answer has sufficient points to gain 4 marks out of 4.

Answer

(ii) The changes have been successful in some ways and unsuccessful in others. The new HYVs have been replaced by even higher yielding varieties (✓).

The problem has been that only the already better off farmers have been able to afford the new HYVs (✓) and machinery and irrigation methods and have therefore increased their yields (✓) but the already poor farmers have been unable to afford (✓) these new things and therefore have become poorer because they cannot compete with the competitive prices (✓) the better off farmers provide. Or they have taken out loans (✓) with extortionate high interest rates (✓) and have been unable to repay them and ended up further in debt (✓).

Comments and marks obtained

The answer contains more than sufficient points to earn pass marks. Points are gained for comments on higher yield varieties being introduced, only affordable by better off farmers as well as machinery/irrigation, increasing their yields, offered at competitive prices which poor farmers cannot compete with, resulting in loans which cannot be repaid causing further debt. This is an upper A grade answer worth 4 marks out of 4.

Impact of tourism in ELDC rural areas

Many ELDCs now make great efforts to attract tourists from wealthier parts of the world. Many thousands of tourists travel to areas such as Mexico, West Indies, Africa, and South East Asia throughout the year.

This industry brings millions of dollars to the economies of these developing countries, helping them to provide jobs, a better infrastructure in terms of roads and railways, improvements to sanitation and water supply and better health facilities for the general population. These improvements often come a cost to the people and the environment.

◆ Coastal and rural areas may suffer from various forms of visual, air, water and other environmental pollution.

♦ Natural environments and habitats of animals may be destroyed to make way for tourist developments.

♦ The cost of living may also rise for local people who remain poor.

♦ Only the main tourist areas may benefit from improvements to the economy and infrastructure.

♦ Other areas may not improve, so cities may still have shanty town housing and poor levels of water and sanitation provision.

♦ Beaches may become over crowded and local vegetation may be destroyed to make way for hotels and other tourist facilities.

♦ Regardless of changes, many rural areas in ELDCs remain in danger from natural disasters such as volcanic eruptions, earthquakes, tsunamis, hurricanes and floods. When these occur they can result in mass destruction and the loss of thousands of lives, such as in the Asian tsunami in 2004. These are dangers however that many tourists are prepared to ignore when visiting relatively cheaper tourist areas in developing countries.

Questions and Answers

Question 2.10

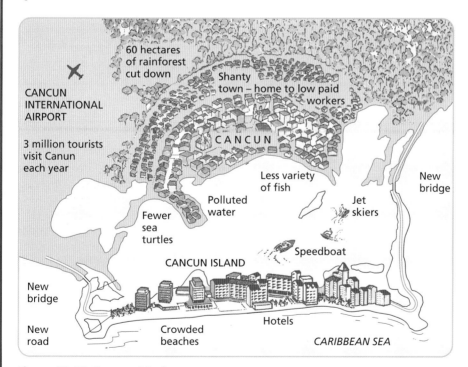

Figure Q2.10 Cancun, Mexico

'Tourism has had a disastrous impact on the area shown in Figure Q2.10'

Questions and Answers continued ➤

Questions and Answers continued

Study Figure Q2.10 and the statement above.

Do you agree with the above statement? Give reasons for your answer. *(4 marks)*

Answer

I agree with the statement above in that tourism has a disastrous impact on the area in the way that the water is now polluted, meaning that there is a lower variety of fish and fewer sea turtles because of the speed boats and jet skiers that the tourists use in the area (✓).

However 3 million tourists visit Cancun each year crowding the beaches and staying in the hotels there (✓). This all provides money for Cancun which can be put back into the area (✓) in the form of new bridges and a new road. Tourism is paying to help improve the area (✓).

Comments and marks obtained

The opening paragraph/sentence gains a mark for noting the pollution of the sea due to the speedboats and jet skiers. Most of the statement is lifted directly from the resource and does not gain any further marks. The next mark is obtained from the reference to crowded beaches indicating a further problem with tourism. The answer then looks at some of the benefits of tourism by referring to money earned for the area from tourism and a final mark showing how that money is used to improve the area. The answer therefore has sufficient good points to gain full **4 marks out of 4**.

Change in Industrial Areas

Key Idea 8

Referring to an area you have studied you should know about the main causes and impact of industrial change.

Depending on the area you have studied you may refer to the fact that many areas throughout Britain and Europe have experienced considerable change during the last 50 years.

Causes of industrial change

A large number of the older industries which provided the basis for economic prosperity have declined due to a number of factors.

◆ Exhaustion of raw materials such as coal, iron ore, and other metal ores.

◆ Loss of markets for products such as ships, iron and steel, textiles and engineering products.

◆ Increased competition from other areas with lower labour costs such as South Korea.

◆ Changing consumer tastes (such as synthetic fibres replacing natural fibres like cotton and wool).

- Increased use of new technology which older industries have failed to develop.
- Changes in government policies which has led to the withdrawal of financial assistance to struggling industries.
- Part of the strategy of allowing older, less profitable industries to decline included replacing them with new sunrise industries such as electronics and other high-technology based industries.

Key Point 26

You should be able to discuss the factors which affect the changing location of industry.

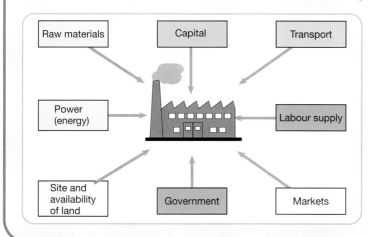

Figure 2.20 **Factors influencing location of industry**

The location of any industry depends greatly on the interaction of several different factors. Each of these factors can influence to greater or lesser extent where a particular industrial unit may be sited.

Raw materials
- Raw materials are the materials from which products are made. They may consist of basic materials such as coal, or semi-finished materials such as strips of steel or rolls of cotton or wool.
- Some industries have to be located at the source of these materials, such as extractive industries.
- Others industries may have to be located very close to their sources of material in order to reduce the cost and difficulties of transport. Many of the older industries had to locate near their raw materials.

Labour supply
- All industries need labour. Modern industries need to consider whether labour is available in a particular location and more importantly whether labour with appropriate skills is available.
- Often companies may locate in areas where older industries have declined leaving a large unemployed labour force needing work. Many new industries require less labour than was previously the case because of the use the introduction of technology.

Power

◆ In the past, power was provided by either water or steam.

◆ Water to drive machinery, or coal or wood to create heat for steam, often meant that industries had to be very close to either or both of these sources.

◆ Sites close to rivers and coal fields were considered the best place to establish industries such as heavy metal production, textiles, engineering works, food processing works.

Transport

◆ Access to industries is necessary both for obtaining raw materials and for transporting the finished product to the eventual market place whether it is another industry or a shop.

◆ Good access may be provided by either road, rail, water or air. Many industrial sites are located near major roads or motorways, railway lines, in river valleys or at the coast, or near major airports.

◆ The type of transport required greatly depends on the type of product being moved. Heavy bulk products such as coal, metals, wood, and textiles may be moved by road or rail transport.

◆ Lighter products which are more expensive may be moved by air transport. In this case the cost of transport may be only a small fraction of the cost of the product being transported (for example, electronics and computer based products, jewellery, medical drugs).

Market

◆ The market is simply the place where the product is sold or service is delivered. The market may be very local as in the case of retailing industries, or can be worldwide in the case of multinational companies such as the motor vehicle or electronic industry.

◆ Few industries need to be located immediately next to their market.

Government and European Union policies

◆ Governments can attract industries to areas by offering a range of grants, subsidies, rent free accommodation and other financial incentives to major industrial companies.

◆ Through its regional aid schemes the EU can offer a wide variety of incentives to help improve the economy of areas which have problems such as high unemployment, industrial decline and slow economic growth.

Industrial and geographical inertia

◆ When an industry such as an iron and steel complex has grown it may remain in an area although the original location factors such as nearness to raw materials no longer apply. This is called industrial/geographical inertia. This may happen because it could be too costly to close the industry and move it elsewhere.

Economies of large factories

◆ Large factories may find it convenient to locate beside each other in order to share the benefits of cost of land or reduce transport costs if trading goods with each other.

Improved technology

◆ Improved technology such as use of computers can allow industries such as communications and newspapers to locate in a variety of areas and have information transferred from the centre to production sites.

Government policies

Local and national governments can have a major influence on the location of industry. At national level, government can intervene in the process of attracting industries to different parts of the country. There are a number of measures which can be used by governments to influence the location of new industries.

◆ Grants can help firms to go to specified areas (often areas suffering from industrial decline), and subsidies can help firms to construct premises, purchase machinery and pay labour costs.

◆ Assistance with labour costs. This can include contributions to retraining schemes and paying additional premiums on salaries to attract workers.

◆ Tax incentives to companies (by reducing certain company taxes for a number of years) can encourage the company to set up business.

◆ Acting with local government to reduce or remove rents and rates for an agreed set period will assist in the early development of the company.

◆ Contributing to the costs of the infrastructure improvements in the area, such as new road or rail links make an area more attractive for new businesses.

◆ Financial assistance for vulnerable industries during periods of economic recession. This may allow the companies to continue trading within the areas they were originally located.

Reasons for adopting these measures include:

◆ boosting the economies and reducing the unemployment rates of depressed areas;

◆ assisting in the process of re-industrialisation by replacing older, declining industries with a newer, more modern and economically successful industrial base;

◆ decentralising industry from the economically stronger areas of the country in the South East of Britain to weaker areas such as the North East and North West of England where many traditional heavy industries have declined;

◆ attracting foreign investment by encouraging non-European countries to set up a base in Britain thus giving them the opportunity to trade within the European community. In this way both the country and the company will enjoy mutual benefits.

Changing industrial landscapes

Key Point 27

You should be able to discuss the ways in which industrial landscapes have changed during the last 50 years by referring to the impact of specific factors.

Changing infrastructure

◆ In the years since the end of the Second World War, many of Britain's older traditional industries such as shipbuilding, iron and steel, textiles and heavy engineering have gone into economic decline.

◆ Many of the large industrial areas of the cities have become abandoned. Factories have been demolished for safety reasons but the former sites have been left as gap sites.

◆ In many cases these have been bought by property developers who have been able to build and sell a variety of development schemes including new housing, shopping and recreational buildings such as sports centres.

◆ Land near the city centre offers attractive sites for the right development.

◆ Cities which had port and shipbuilding industries often have large areas of dockland which is no longer in use. In Glasgow, Liverpool and London these dockland areas have had major changes and now contain a wide variety of new land uses such as new housing, conference centres and hotels and a variety of shopping, tourist and other service functions. These have been very successful and have changed the environments of these former industrial areas completely.

◆ Brownfield sites of former manufacturing industry have been developed as new industrial and trading estates and science parks.

Appearance

Industrial Zones

◆ As older heavy industries declined during the 1970–1990 period, these were often demolished leaving brownfield sites within cities for new developments such as housing, shopping centres or more modern industrial estates.

◆ Many new industries found sites on the outskirts of cities. These sites were often on cheap land with good access by way of motorways being close by.

Fringe developments

◆ The styles of medium and high cost houses have changed considerably. Many private housing estates have been built on the fringes of cities or in small villages on the outskirts attracting people who are willing to commute to the city on a daily basis for work.

◆ During the 1960s large numbers of people moved from cities to new towns several kilometres away from the older urban areas.

◆ New towns offered new housing and the prospect of new jobs since many modern industries were attracted to them.

◆ New Town councils often offered incentives such as subsidies and rent-free accommodation to attract these industries.

◆ In Britain, new towns such as East Kilbride, Crawley and Milton Keynes were very successful in attracting people and industry away from the older urban centres.

Impact of change on the environment

Key Point 28

For an area you have studied, you should be able to describe and account for recent industrial change and the impact of this on the landscape and how industrial regeneration has affected the area.

Case study
South Wales

During the 1970s the industrial landscape of South Wales gradually began to change. With the help of government financial aid, several new modern factories were encouraged to set up production in various parts of the area. These industries included electronics companies, sweet making factories, shoe making and a variety of other modern industries.

Transport

◆ New industrial estates were built next to motorways such as the M4.

◆ Good access for the factories in these estates for bringing in raw materials and distributing their finished products was essential. This reduced transport costs and improved the efficiency of the system.

◆ Good transport links were also essential for the labour force which staffed the factories.

Government

◆ The Government used financial incentives to attract British and foreign firms to the older industrial areas of South Wales.

◆ These incentives included rent-free accommodation; grants for new machinery; subsidies to assist with labour costs; grants to help set up retraining schemes; and tax incentives to encourage foreign companies such as Sony.

Effects of regeneration

◆ As a result, a wide variety of different manufacturers were attracted to the area.

◆ In addition to these, old closed mines were refurbished and reopened as tourist centres and educational centres.

◆ In the Rhondda valley, houses and mining villages were rebuilt along with theme parks, boating areas and dry ski slopes.

Key Point 29

You should be able to describe the environmental impact of new industrial developments in your chosen area.

For South Wales:

◆ modern buildings replaced old dilapidated buildings;

◆ new road systems replaced older roads easing the flow of transport throughout the area;

◆ new modern industrial estates, planned to enhance rather than spoil the landscape replaced worn out old factory buildings.

The modern industrial environment is now relatively free from pollution. Green field sites are home to new modern light industrial units.

HUMAN ENVIRONMENTS

Geographical methods and techniques

You will be asked to use different geographical methods and techniques to describe and evaluate changing industrial location and landscapes.

Referring to Ordnance Survey maps in questions is very common. The main points which you should look for include location of the industrial area under discussion, trying to identify characteristics of industry present – old or new industry, size of buildings, individual factories or industrial estates.

Other factors may come into your analysis such as the type of transport system available, proximity to work force, type of local road pattern, access to local universities for research, airports or ports if they are shown on the map.

If you can recall the main factors affecting the location of industries and can apply them to specific situations, you should be able to tackle most map-based questions.

You will not be asked to annotate diagrams in the external exam but you do require this skill to attempt National Assessment Bank questions in the classroom.

Examination questions based on an OS map extract will ask you to explain the location of a particular industry using map evidence.

◆ If you can remember the main location factors you should inspect the map for suitable evidence to support your explanation. This may include reference to flat land (absence of contours), nearness to raw materials, accessibility due to the proximity of road and rail networks, being near a source of labour such as a town or city, and room to expand and develop.

◆ Other questions may provide you with information on a map or diagram and ask you to evaluate the site of an industry in terms of suggesting the main advantages and disadvantages.

◆ In answering these questions make full use of the information given to support your answer. You will be given full credit for carefully selecting appropriate information from the given source.

Analysis of OS Maps

◆ Many of the questions asked in the unit tests and the external examination concentrate on the interpretation of a variety of maps which contain different examples of industrial land use.

◆ Essentially these questions ask candidates first to describe both the distribution and type of industries present, and second to explain their distribution and location pattern.

◆ You need to correctly identify the type of industry present. The map might contain old, traditional industries, extractive industries, manufacturing and service industries, new modern light industry and industry which is transport or port related, such as shipbuilding, dockyards, warehouses, railway stockyards or aircraft industries.

Figure 2.21 illustrates three different industrial areas which could be seen on an OS map.

Figure 2.21a shows a site of primary industry

Primary industries such as mining, quarrying, forestry estates and farming should be easily recognised on OS maps. Their location is obviously determined by the presence of the raw material being extracted and in the case of farming, the height and shape of the land.

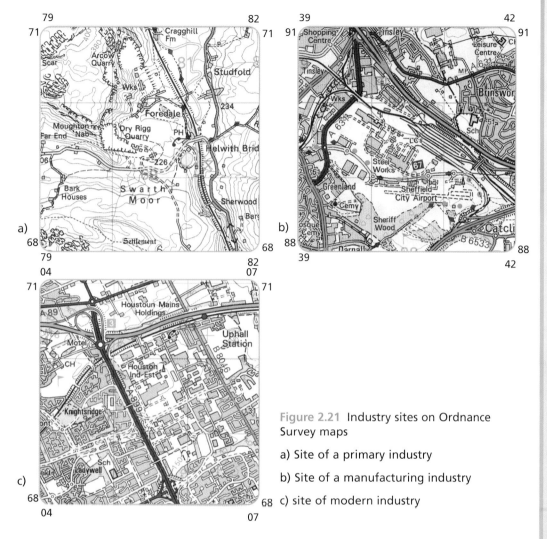

Figure 2.21 Industry sites on Ordnance Survey maps

a) Site of a primary industry

b) Site of a manufacturing industry

c) site of modern industry

Occasionally there may be examples of some former extractive industry and the prefix *dis* will be seen. This indicates a disused mine or a disused quarry and is often a good guide to the changing pattern of industry in an area.

Figure 2.21b shows a site of a manufacturing industry

Large manufacturing industries such as iron and steelworks or engineering plants should be easily spotted by the shape of the buildings on the map as shown on figure 2.21b.

Often the word *works* will be written beside the factory with other descriptive words such as *aluminium*. Power stations are also easily identified by written description beside them. Similar written descriptions may accompany *mills* and *distilleries*. Often large factories may have their own railway line leading into the factory from the main line.

Most older industries in a city will be located close to the centre of the city, perhaps surrounding the central business area. Most will be near the main road and rail arteries leading to the centre.

Many industries may be located along one or both banks of a river including port industries, docks, shipyards, warehouses, food processing plants and petrochemicals, oil refineries and possibly even iron and steel factories. The sites of former industries are often used as development sites for newer industrial units since the land may be cheaper and more easily obtained than in other parts of the settlement.

Figure 2.21c shows the site of modern industry

Modern industries will generally be found in industrial estates either within or on the boundary of large urban areas, and are shown by the words *ind. est.* on the map. Buildings will be smaller and laid out in a more planned manner than the larger factories. Nearby street patterns will probably have less of a grid pattern than those with the older industries and there may be easy access to a nearby motorway.

A more recent development has been the emergence of science parks and enterprise zones or trading estates in many settlements. These may be named on the map and the science park may be named after a local university.

Questions and Answers

Question 2.11

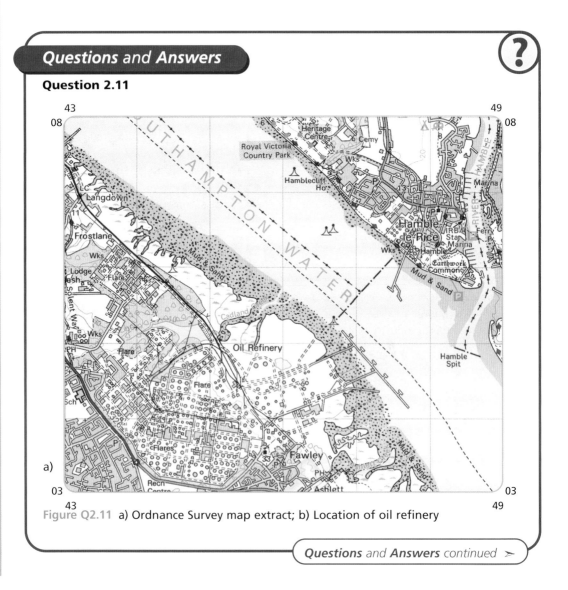

Figure Q2.11 a) Ordnance Survey map extract; b) Location of oil refinery

Questions and *Answers continued* ➤

Questions *and* Answers *continued*

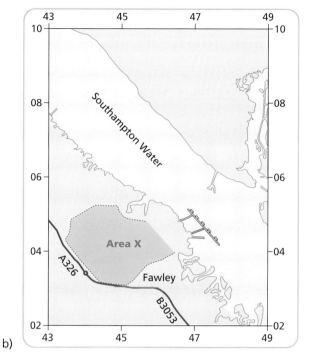

b)

Study Figure Q2.11. Explain the suitability of Area X for the location of an oil refinery. *(4 marks)*

Answer

The suitability of area x for the location of an oil refinery is quite good because it is a large area of flat land suitable for building on (✔) with good communication links nearby such as the railway line (✔) can be seen in grid square 4405. There is also the B3053 and the A326 close by which would provide a close link to transport materials (✔). Being close to the water means they could also import and export their goods by sea (✔). However it is this very point which can make an oil refinery unsuitable for this area as if there was an accident at the refinery, oil could spill out into the sea, polluting it (✔).

However the main point it is suitable particular as it is near to towns such as Fawley which could provide a workforce (✔) for the refinery.

Comments and marks obtained

This is an excellent answer which provides good evidence from the OS map to support points made. The answer obtains marks for noting the flatness of the land – suitable for building on, transport links, naming both road and rail. It also gains a further map for reference to sea transport for imports and exports. A further mark is obtained for the good reference to sea pollution. A final sixth mark would have been given for the comment on nearby workforce. The answer has six good points and gains a maximum of **4 marks out of 4.**

HOW TO PASS INTERMEDIATE 2 GEOGRAPHY

Glossary *Human Environments*

Active population: That section of the population of a country which is economically active and working.

Agribusiness: The operation of a large scale farm which resembles a factory with large investments in farm property, maintenance, machinery and technology.

Arable farming: Farming where the main activity and income source is the growing of crops.

Birth rate: The number of births per thousand of the population in any country in any year.

Brownfield site: Land in an urban area on which development (often in the form of factories) has previously taken place.

Business Parks: Areas which have become industrial estates with businesses which may sell products or provide services directly to the public.

Casual workers: Workers employed at specific times during the year (such as strawberry pickers in the summer).

Census: A numerical count of the population financed and carried out by the government at set periods of time (10 year intervals in the UK).

Central Business District (CBD): The zone which contains the major shops, businesses, offices, restaurants, clubs and other entertainments and is normally located at the centre of the settlement at the junction of the main roads.

Cereal crops: Grain crops such as oats, barley or wheat.

Common Agricultural Policy: A system used by the European Union to give farmers guaranteed prices for their products.

Commuter settlement: Small settlements on the outskirts of major towns and cities where people live but travel into the main settlement for employment and services.

Commuters: Those who live in commuter settlements.

Crofting: A type of mixed farming found in northern Scotland which is not very profitable. Crofters often have to supplement their income by doing other part-time jobs.

Crop rotation: A system designed to maintain the fertility of soil by growing crops in different fields from time to time.

Dairying: Farming in which the produce is derived from milk. Cows are reared to supply milk on a daily basis. These have become highly mechanised with milking machines, high standards of hygiene and milk lorries take the milk away to dairies to be processed into various products. Farmers are paid on the amount of milk supplied.

Death rate: The number of deaths per thousand of the population of any country in any given year.

Dereliction and decay: This refers to closed and abandoned buildings such as mines, offices or industries. They often become a source of visual pollution if they are not demolished.

Glossary   *Human Environments continued*

Developed countries: Sometimes referred to as Economically More Developed Countries. These are countries which have a high standard of living or high quality of life.

Developing countries: Often referred to as Economically Less Developed Countries in which the population generally has a low standard of living.

Diversification: The process whereby an economic enterprise such as a farm takes on a range of additional activities to increase profits for example, renting land for golf courses, caravan sites, paint ball enterprises, quad biking.

Drainage: If the underlying rock is clay, bogland and marshland may develop. Pipes are laid to drain excess water from the surface and allow the land to be farmed.

Economic: Relating to financial developments.

Economic effects: The financial impact of change on for example employment, incomes, running costs, building costs and costs to the local community.

Empty lands: Areas of the world which have a low population density, such as mountains and deserts.

Enterprise zone: An area which receives government assistance to attract new industry and create new employment opportunity.

Environmental consequences: The effects of change on the physical and human environments for example changing land use, improvements and bad effects on the environment such as increased pollution, changes to the population caused by people moving to or from areas as a result of change such as industrial closures or new industry being built.

Environmental factors: Factors such as climate, relief, soil and water supply which can influence the distribution of population in an area.

Extractive industry: Primary industry which takes raw materials from the ground such as mining, quarrying and drilling for oil and gas.

Farm system: The relationship between inputs, processes and outputs on a farm.

Fertilisers: Substances which may be organic or chemical and are added to soil to increase fertility and improve crop yields.

Fodder crops: Crops grown on a farm to feed animals, like grass and turnips.

Functions: These are individual activities which settlements perform such as commercial, industrial, administrative, transport, religious, medical, recreational and residential.

Functional zones: These are areas of a settlement where certain economic or social functions are dominant.

Green belts: Areas surrounding cities and towns in which laws control development such as housing and industry in order to protect the countryside.

Greenfield site: Land which has not been previously used for industry or any other buildings.

Glossary   *Human Environments continued*

Gross Domestic Product (GDP): The value of all goods and services of a country produced in one year and is used as an indicator of the wealth of a country. However it does not always reveal how well spread the wealth is among the population in general.

Heavy industry: Industry which produces heavy bulk materials such as iron and steel, shipbuilding, stone, clay, and glass products and cement.

High order functions: Functions considered to have the highest economic or social value such as department stores, Council offices or art galleries.

High tech industry: Industry which uses the latest technology to produce goods and services.

Industrial decline: Closure or reduction in the scale of an industry sector. In the UK this has involved traditional, heavy industry such as mining, iron and steel making, textiles and shipbuilding, largely as a result of the easy availability of cheaper imported goods.

Industrial estate: An area set aside for modern, light industrial units, often located on the outskirts of towns and cities where land is available and the area is served by a good communication system such as motorways.

Industrial growth: Increase and development of industrial output. This often results in increased number of workers and profits and usually has a positive effect on the local community.

Industrial inertia: This occurs when an industry remains in an area long after the original location factors no longer apply.

Infant mortality: The number of children below the age of one year which die per thousand of the population.

Inner city: The area near the centre which contains the CBD, the older manufacturing zone and zone of low cost housing.

Inputs: The basic needs of a farm before the farmer can begin to farm the land such as seeds, livestock and machinery.

Insecticides: Chemicals sprayed on to crops to kill insects which may be attacking crops and therefore destroying the yield.

Life expectancy: The average age a person can expect to live in any given country. This is a good indicator of level of development, since people in more developed countries tend to live longer due to better health care, better diets, higher standards of education and housing etc.

Light industry: Industries which manufacture small, light bulk products (like window frames).

Low order functions: Functions considered to have low economic or social value such as small corner shops, post offices and petrol stations.

Manufacturing industry: Industrial activity which actually makes finished or semi-finished products.

Glossary   *Human Environments continued*

Multiplier effect: The effect of change on other activities such as other industries, settlements or rural activities. For example when a major industry closes this may cause local shops and other business to close since people made unemployed have less money to spend and may move elsewhere to seek new work.

Population density: The average number of people within a given area. Usually expressed as a number per square kilometre.

Population structure: The grouping of the population of a country by age and sex. The structure may also indicate trends in birth and death rates, life expectancy and the possible impact of factors such as war and migration on the population.

Mixed farms: Farms based on a combination of arable and pastoral farming.

Outputs: The end product from the inputs and the processes of production on the farm.

Overspill population: People who have moved out of the main city to other smaller towns or new towns.

Park and Ride schemes: Schemes which allow people to park their cars outside city limits and to use an integrated public transport system for the remainder of their journey into the city. This is an attempt to reduce traffic congestion.

Pastoral farm: Farms based on rearing livestock such as beef cattle and upland sheep farming.

Pasture: Land grazed by livestock. Some of this may be permanent, some temporary and some might be poor pasture which has been improved by drainage schemes.

Pedestrianised zones: These are traffic-free areas within the city centre where people can shop and walk along streets where traffic is banned.

Population Growth Model: A theoretical model to show different stages of population growth based on the relationship between birth and death rates.

Primary industry: Industries involved in converting natural resources into primary products, often producing raw materials for other industries. Major businesses in this sector include agriculture, fishing, forestry, mining and quarrying industries.

Processes: Work done on the farm and obviously varies according to the type of farm.

Quotas: Limits imposed on farmers in order to limit the output of certain types of produce to avoid surpluses and therefore a drop in prices.

Renewal and regeneration: The processes by which older areas are demolished and replaced by new buildings often having totally different functions to the original building or area.

Ring roads: Roads built specifically to take traffic away from the city centre and to help solve the problem of congestion.

Root crops: Crops harvested for their roots such as potatoes, turnips and carrots.

Glossary   *Human Environments continued*

Rough grazing: Poor quality land used for grazing for example upland areas where soil is thin and is grazed by sheep.

Site: Land upon which a settlement was originally built.

Site factors: Factors which influence people to choose a particular site. Factors can include nearness to water supply (river), flat land for building, high land for defence, at a suitable point on a river where a bridge could be built, near raw materials, etc.

Science parks: Industrial areas closely connected to technological institutions and universities. They are often involved in producing high technology products and have often helped to replace declining industries by providing new job opportunities to workers from closed industries following retraining schemes.

Service industry: Also referred to as tertiary industries which provide services such as retailing, wholesaling, banking and finance, transport, legal, and administration.

Sphere of influence: The area from which settlements draw its customers for various functions. Usually the size of the sphere of influence varies in direct proportion to the size of the settlement.

Standard of living: The level of economic well-being of people in a country.

Suburbs: Housing zones on the outskirts of towns away from the busy central inner city.

Sunrise industries: New highly technological industries such as electronics. These industries are normally growing and gradually replacing older, declining industries.

Traffic congestion: The heavy build up of traffic along major routes and within city centres. Congestion causes great problems of cost and pollution for many cities in the UK.

Urban areas: Another term for cities and towns.

Chapter 3

ENVIRONMENTAL INTERACTIONS

Section A
Rural Land Degradation

The regional context in which this topic is studied is any area in the world except the British Isles.

Key Idea 1

You should know about two examples of land degradation. The two specific types of land degradation are deforestation and desertification related to arid and semi-arid areas.

Key Point 3

With specific reference to deforestation and desertification, you need to know about general features of the physical environments such as climate, vegetation, soils and relief.

Key Point 3

You also need knowledge of features of the human environments such as population density and farming systems. Knowledge of these features should be acquired during the course of your work in class.

This section of the book provides a review of the causes, effects of degradation and management strategies to deal with this problem and their effectiveness.

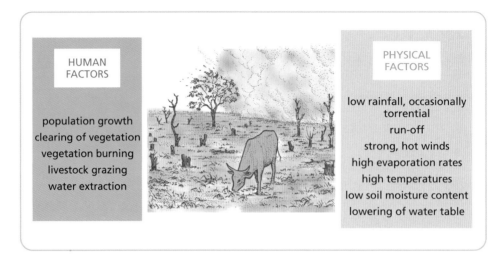

HUMAN FACTORS

population growth
clearing of vegetation
vegetation burning
livestock grazing
water extraction

PHYSICAL FACTORS

low rainfall, occasionally torrential
run-off
strong, hot winds
high evaporation rates
high temperatures
low soil moisture content
lowering of water table

Figure 3.1.1 Causes of land degradation

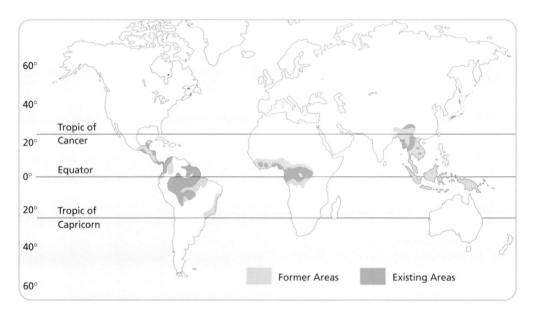

Figure 3.1.2 Deforestation around the world

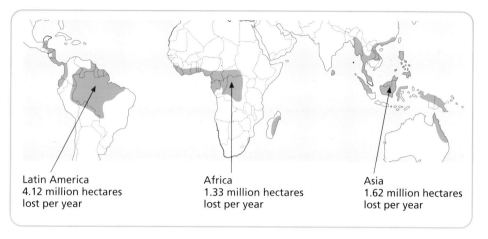

Figure 3.1.3 **Annual destruction of rainforests in selected areas**

Tropical rainforests

Tropical rainforests are located in the land areas between the Tropics of Cancer and Capricorn in all continents. The climate has high temperatures and high rainfall throughout the year as shown in Figure 3.1.4.

Physical features of tropical rainforests
◆ Vegetation consists mainly of hardwood trees such as teak and mahogany rising to over 50 metres and forming a canopy of several layers.

◆ A wide variety of plant species and wildlife exists within the forest.

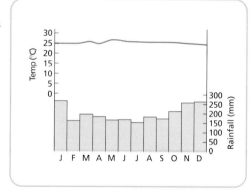

Figure 3.1.4 **Graph of equatorial climate**

◆ Soils include forest soils which, when exposed to rainfall and sun through the removal of trees, become leached. In this process minerals rise through the soil to the surface forming what is termed a hardpan. When this happens the soil becomes infertile.

Human features
◆ Population density is generally low within the forests and consists largely of indigenous tribespeople, communities of farmers, rubber tappers and miners.

◆ The main farming systems that exist include subsistence shifting cultivation, large scale ranching and small-scale crop farming.

◆ Shifting cultivation is the traditional type of farming practiced by tribespeople. The system is based on small groups of Indians who clear a small part of the forest and over a period of about 5 to 8 years cultivate small patches of land. Crops include yams and manioc. The diet is supplemented with fish from local rivers, meat from animals hunted in the forest and local forest fruits gathered by the tribes.

◆ Ranching is the tending of cattle on large areas of land which has been made available through large-scale destruction of the forest.

◆ Land has also been made available by governments for small-scale farmers, many of whom have been encouraged to migrate from cities to rural areas.

Throughout equatorial areas of the world thousands of square kilometres of trees have been destroyed during the last 50 years in various ways and for a variety of reasons.

Methods of destruction include:

◆ destruction by fire (often the easiest and quickest way to destroy forests);

◆ cutting down trees for logging;

◆ drowning due to valleys being flooded to create reservoirs for multi-purpose water schemes.

Despite the damaging effects on the environment, people in countries such as Brazil and the Congo and parts of India and South East Asia continue to destroy the forest.

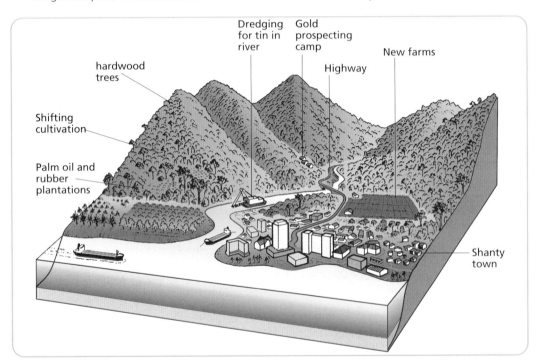

Figure 3.1.5 **Developing the tropical forest**

Reasons for the destruction of rainforests

◆ Cutting down hardwood trees for timber for export. Hardwoods such as teak and mahogany can be sold for large sums of money.

◆ Clearing areas for small-scale farmers and to create grazing land for cattle.

◆ Clearing trees for mining of minerals such as iron ore, copper, bauxite and gold.

◆ Using trees for fuel such as charcoal for local industry such as iron smelting.

◆ Flooding valleys to create reservoirs for hydroelectric power schemes.

All of these reasons are aimed at obtaining money from destructive activities.

The effects of deforestation

Destruction of trees can have a devastating effect on people and their environment.

◆ Rivers become polluted from mining enterprises, killing fish and affecting food supply and welfare of tribes living in the forest.

◆ Without trees to interrupt run-off of rainwater, serious flooding can occur causing widespread death and disaster.

◆ Thousands of different species of animals, insects and plants are destroyed each year.

◆ Deforestation reduces the amount of carbon dioxide taken in from the atmosphere. This is one factor in the increase in levels of carbon dioxide in the atmosphere, which is contributing to global warming.

Measures to reduce deforestation and their effectiveness

◆ Laws have been introduced to protect forests, by limiting the amount of land which can be used for activities such as mining and ranching. These laws are often very difficult to enforce.

◆ Worldwide campaigns by various protest groups such as Greenpeace are aimed at saving the forests.

◆ Encouragement of other commercial developments within the forests by buying forest products such as tropical fruits.

Unfortunately the process of deforestation is continuing faster than the efforts to reduce deforestation.

There are many who live in the forest who can work in harmony with the forest including tribes who hunt, gather and are subsistence farmers, rubber trappers, farmers and those who replant areas which have been cut down with young trees.

Case study
Amazon Basin

Impact on people and social and economic consequences

◆ Traditional ways of life are damaged and destroyed, with clashes between locals and incomers. Formerly sustainable activities such as rubber tapping are destroyed.

◆ Intensive farming reduces fallow periods for land, causing reduced crop yields and food shortages.

◆ Large business interests often act in conflict with local interests.

◆ Migration of local people away from traditional habitats is increasing poverty and social deprivation.

Impact on land and environmental consequences

◆ Clearing of vegetation causes leaching and laterisation of soils exposed to the elements, disrupting the nutrient cycle and reducing the land's productivity.

◆ There is increased run-off of ground water and flooding.

◆ There is a loss of wildlife habitats.

◆ Flooding due to deforestion can remove soil and mining, causing increased pollution.

◆ There may be an impact on local climate due to lack of moisture recycling.

◆ There are wider effects on global climate through the greenhouse effect.

Strategies to manage and reduce deforestation in the Amazon Basin include:

◆ reforestation with mixed trees;

◆ the use of crop rotation by farmers;

◆ the purchase of forest areas by conservation groups;

◆ returning forests to native peoples.

Several schemes are very effective but outside interests in mining and ranching often take precedence over conservation measures. There have been attempts to control this through government legislation but the impact has been limited to economic demands for development.

Questions *and* Answers

Question 3.1.1

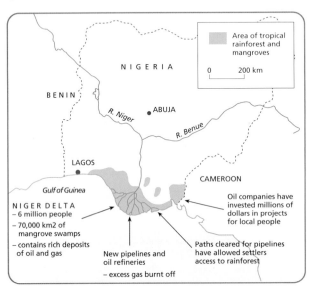

Figure Q3.1.1 Oil industry in the Niger Delta

Table Q3.1.1 Facts about oil in the Niger Delta

% of Nigeria's exports provided by oil	90
Number of oil spills in the Niger Delta 1976–1991	2976
Number of people employed in the oil industry in Nigeria	25 000

Study Figure Q3.1.1 and Table Q3.1.1

The damage caused to the rainforest environment in the Niger Delta is a small price to pay for the huge benefits which the oil industry has brought to the people of the area.

Do you agree with this statement? Give reasons for your answer. *(6 marks)*

Questions and Answers continued ➤

Questions and Answers *continued*

Answer

I agree that there are huge benefits from the oil industry but the rainforests should not have to be cut down.

The oil companies have invested lots of money in projects for local people which gives jobs (✓) and therefore money to buy food and clothes and proper houses (✓). This is an excellent idea for these people. The number of people employed in the oil industry is 25 000, which is great for the country (✓). The oil provides 90% of Nigeria's exports as well (✓).

Despite all of this the rainforests are being destroyed because of it. Many people's houses and habitats are being destroyed with the forests (✓). The new pipelines are built and thousands of people and wildlife lose their habitats because of it. (✓). There have been 2976 oil spills between 1976 and 1991 and nobody knows the amount of wildlife this has killed (✓). The trees provide the world with vital vitamins and minerals and other chemicals which are used to make medicines (✓).

Comments and marks obtained

This is a very good example of an answer which makes good use of the resources given in the question. It also follows the instructions in the question, namely to discuss the advantages and disadvantages of the developments in the rainforest. The answer obtains additional marks by including relevant pieces of data from the question such as percentages of exports, the number of oil spills and relates this to good and bad effects of the development. The first part of the answer has sufficient points to gain full marks and the second section also has sufficient detail to gain a further full marks. Therefore the answer merits a total of **6 marks out of 6.**

Question 3.1.2

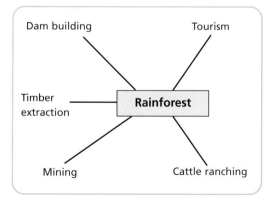

Figure Q3.1.2 Exploiting the rain forest

For an area you have studied, explain the impact of these activities on the people and landscape. *(6 marks)*

Questions *and* Answers *continued* ➤

Questions and Answers continued

Answer

In the Amazon rainforest, trees have been cut down for wood supplies and to make way for mining and cattle ranching. This has meant that animals and plants have been destroyed (✓) as have the habitats of many species of animal (✓). Local tribespeople have lost their homes and have been forced to move elsewhere (✓). The loss of trees has meant that there are no trees to protect the soil from erosion from heavy rainfall (✓) and there are no roots to bind the soil together. Soil is washed away causing erosion (✓) and is deposited in rivers, polluting local rivers (✓). This kills fish (✓) and local people lose a source of food supply (✓). Destroying the trees affects the atmosphere since trees give off oxygen to the atmosphere (✓). Burning trees adds carbon dioxide to the air causing pollution which can lead to global warning (✓).

Comments and marks obtained

This answer contains many good points about the impact of rainforest destruction. The answer refers to the Amazon rainforest and therefore qualifies for full marks. The reference to destruction of plants/animals and habitats merits two marks. A further mark is obtained for reference to local people being forced to move. Further marks are gained for the references to soil erosion, pollution of rivers and the loss of food supply. Finally, marks are gained for the mention of carbon dioxide and global warming. The answer has covered both impacts on people and landscape and obtains **full six marks**.

Deserts

Physical features

◆ Not all desert areas consist of sand. Many consist of dry rocks and tracts of sand.

◆ The climate consists of high temperatures throughout the year, in some cases in excess of 40°C and low rainfall, often resulting in long periods of drought. Figure 3.1.7 shows a typical desert climate graph.

◆ Soils are known as desert soils. They are generally infertile, due to the lack of moisture and nutrients.

◆ Whenever rainfall does occur it often results in flash floods, which wash away topsoil, leaving the land even more infertile.

◆ Areas within deserts are characterised by dried-up river beds (wadis).

◆ Oases provide the main natural water source in deserts.

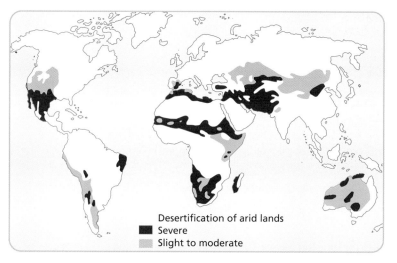

Figure 3.1.6 Global desertification

Human features

◆ Population density is generally very low apart from the settlements which are based around oases.

◆ The main type of farming is pastoral nomadism. This involves tribesmen who travel with their livestock throughout the year in search of new pastures for their animals, (mainly sheep and goats).

◆ Where water is made available through irrigation, crop farming is possible. Crops include wheat, maize, cotton and tobacco. This farming is usually found in the floodplains of rivers such as the Nile.

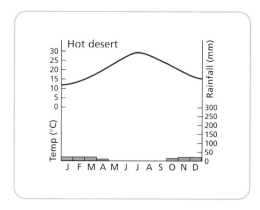

Figure 3.1.7 Desert climate

Desertification

Desertification is the process by which formerly productive land is turned into desert. At the edges of hot deserts throughout the world, the deserts are spreading into areas which were formerly settled and farmed and provided a living for many people. Although there are physical factors responsible for this process including prolonged periods of drought, human factors are probably most responsible for this process occurring. The impact of many human activities has led to land turning into deserts.

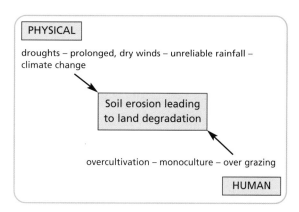

Figure 3.1.8 Causes of land degradation in arid and semi arid areas

Causes of desertification

◆ The main physical cause of desertification is prolonged absence of rainfall, leading to prolonged drought.

◆ As a result plants and vegetation die and there is little cover over soils. Wind erosion strips away dead vegetation and top soil, eroding the land and depositing fine sand elsewhere.

◆ Insects may also contribute by eating vegetation and removing soil protection.

On top of these natural processes, human activities can help accelerate the process of desertification. These activities include:

◆ deforestation by which trees are cut down for firewood;

◆ farmers allowing their animals such as sheep and goats to overgraze the vegetation leading to more soil erosion;

◆ farmers over-cultivating the land removing any goodness in the soil. This reduces the ability of the soil to support any more crops.

People carry out these activities because of the harshness of the environments in which they live and the lack of alternatives. There is a desperate need for food during periods of extensive drought and a lack of expertise in proper farming techniques.

Effects of desertification

◆ Previously fertile farmed land is abandoned. Without preventative actions, the abandoned land turns to desert, increasing the extent of deserts.

◆ Farmers cannot graze animals on desert lands.

◆ Soils are exposed to the elements (sun and wind) and are eroded and are unable to sustain crop growth. Famine occurs due to low crop yields

◆ Since the land cannot sustain any further farming, local people have little option but to move away to areas where the soil might be more fertile. Villages are abandoned. People may migrate to other areas. Traditional farming methods such as pastoral nomadism may disappear.

Management strategies to reduce desertification and their effectiveness

◆ The most effective measure which can be taken to reduce desertification is to bring water to dry areas. This can be done through various forms of irrigation, such as large scale multi-purpose water projects to bring water to dried-up land, and smaller-scale local projects digging wells and irrigation ditches.

◆ Governments can help farmers by providing people to advise them on better farming methods such as using fertilisers and 'miracle seeds' where possible.

◆ Soil conservation methods are used, such as fencing off grassland areas to avoid overgrazing and planting young trees to act as windbreaks.

◆ Financial assistance is provided to governments through international aid agencies such as World Health Organisation, United Nations Food and Agricultural Organisation and the World Bank, and loans and aid programmes from more developed countries throughout the world.

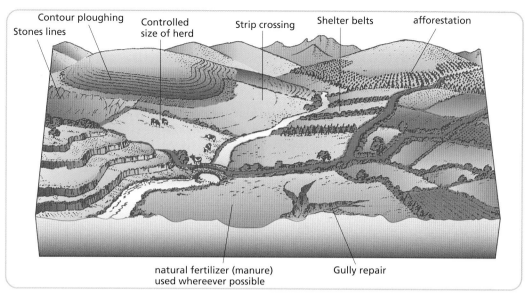

Figure 3.1.9 Conservation measures to prevent desertification

Case study
Africa, north of Equator

Social and economic consequences of desertification

◆ Crop failures and consequent malnutrition lead to major famines (seen recently in Ethiopia and Sudan). This leads to mass migrations to refugee camps or shanty towns on edge of cities.

◆ There is often a collapse of traditional activities such as nomadism due to overgrazing and lack of water.

◆ There is increased pressure on land due to nomads settling in villages. This results in increased tensions between nomads and traditional farmers.

◆ There may be widespread desperate poverty and deprivation. This results in increased mortality rates, and especially infant mortality rates.

Impact on land and environmental consequences

◆ There is a breakdown of soil structure, with wind and rain erosion of dried-out soil. This causes an advance of the Sahara desert resulting in the process of desertification.

◆ Water tables are lowered.

For Africa north of the Equator, solutions include:

◆ building dams (as is happening in parts of Kenya);

◆ planting trees as windbreaks, terracing slopes to prevent erosion, and stabilising sand dunes with grass (as is happening in parts of Mali);

◆ improving irrigation (even small-scale schemes are very effective in increasing soil depth and crop yields);

◆ controlling grazing and fencing, which is effective in preventing the removal of vegetation, and reduces soil erosion due to wind action.

Examination hints

If you are asked in the examination for either an opinion on the different causes for these types of land degradation, or alternatively to judge the main advantages or disadvantages of perhaps different strategies to manage the problem, you will be provided with a resource such as a map or diagram.

You must be able to use the information given in the source to support points which you make in your answer. Marks are gained for selecting the appropriate data in support of your opinions or choice. The main difference in your answers between Intermediate levels 1 and 2 is the amount of detail which you need to provide in your answer. Use the number of marks allotted to the question as a guide to the depth and amount of detail required.

Questions *and* Answers

Question 3.1.3

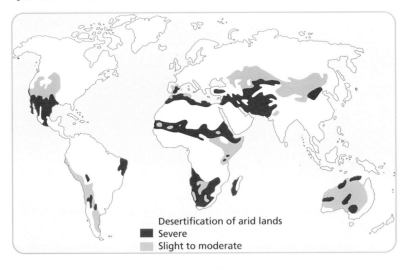

Desertification of arid lands
■ Severe
▨ Slight to moderate

Figure Q3.1.3
Desertification

What are the main causes of desertification?   *(4 marks)*

Intermediate 2

Answer

When you clear away trees, there are no longer any roots to hold the top soil together (✓) so this blows away causing desertification (✓). Overfarming whereby not allowing any fields to lie fallow (✓) and constantly grazing animals (✓) and growing crops on the same land every year drains the soil of nutrients (✓) so again it blows away causing desertification.

 Comments and marks obtained

Although this answer does not refer to some of the main causes of desertification such as prolonged periods without rain i.e. drought, nevertheless by referring to tree removal and its effects and overfarming the land, giving good examples of how this happens, the answer does gain enough marks to award more than the four marks available. It therefore gains 4 out of 4 marks.

Glossary *Rural Land Degradation*

Cattle ranching: Rearing of large herds of cattle on areas of cleared forest to provide beef to be sold for export.

Deforestation: Removal of trees usually on a large scale.

Degradation: The process of reducing land which was formerly productive into unproductive land.

Desertification: Process of turning land which was formerly productive into desert.

Drought: Prolonged periods without rainfall which may last from several months to several years.

Food chain: A system whereby various forms of life provide food for each other starting at one point finishing at another e.g. from small fish to larger and eventually to humans.

Global warming: The heating up of the Earth's atmosphere by the sun's rays due to the presence of greenhouse gases in the atmosphere.

Greenhouse effect: The warming effect due to the presence of gases like carbon dioxide and methane in the atmosphere. Increased levels of these gases as a result of burning fossil fuels (coal, oil and gas) and as a result of deforestation is leading to global warming and climate change.

Greenpeace: An international organisation which campaigns against the causes of world pollution and other elements which destroy the natural environment.

Insecticides: Chemicals used by farmers to kill insects which feed on crops.

Irrigation: An artificial way of providing water for farming from sources such as rivers, wells, canals, field sprays.

Logging: A commercial business which cuts down trees to provide timber for sale for different purposes.

Marginal land: Land on the outer limit of sustainable growth and development where crop growth is only just possible.

Nomadism: The system whereby people migrate with animals throughout the year to find new areas for grazing livestock also known as pastoral nomadism.

Overcropping: Growing crops continually to the extent that the soil becomes exhausted of nutrients, becomes infertile and unable to sustain further growth.

Overgrazing: Allowing animals to overeat grass to the extent that the underlying soil is exposed and cannot sustain further growth.

Pollutants: Material which is released into the environment which ultimately causes damage to the physical landscape and atmosphere.

Power stations: Complexes which are built to provide electricity from various sources such as nuclear energy, fossil fuels and water.

Reforestation: The process of replanting trees in former forested areas.

Glossary *Rural Land Degradation continued*

Shelter belt: A line of trees planted to provide shelter from the wind for fields by interrupting the flow of wind.

Soil conservation: Attempts to protect soil from damage using methods such as fertilisers, irrigation, shelter belts and ploughing along contours and using terraces to conserve soil.

Soil erosion: The process by which the top soil is removed leaving the land infertile.

Windbreaks: Trees which interrupt the flow of wind so as to protect fields and their crops.

Section B
River Basin Management

Key Idea 1

Water control projects are undertaken within river basins for various reasons.
Changes brought about by human interference with river systems can have both
beneficial and adverse consequences. You should have detailed knowledge of a river
basin located outside Europe.

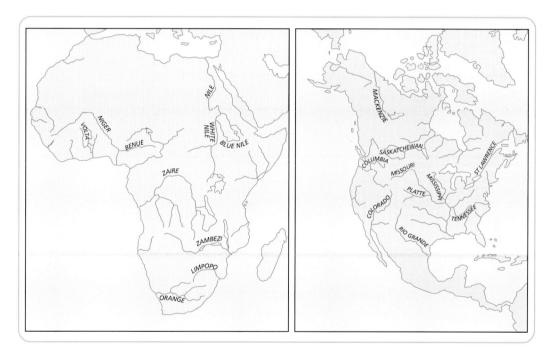

Figure 3.2.1 Major river basins in Africa and North America

ENVIRONMENTAL INTERACTIONS

Key Point 1

You should know the general components of the hydrological cycle, shown in Figure 3.2.2.

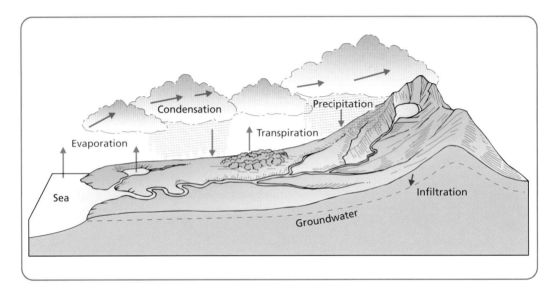

Figure 3.2.2 **The hydrological cycle**

◆ Water exists on the surface in the form of oceans, seas, lakes, rivers and streams. Water also exists on the surface as ice or snow at the Poles or in high altitudes.

◆ Water exists in the atmosphere as rain and water vapour and underground as seepage within rock structures and in underground streams and lakes.

◆ The surface water passes into the atmosphere through evaporation and is carried by winds and eventually returns to the surface as rain or snow.

◆ Water on the land may be returned to oceans and seas through rivers or streams.

◆ This intricate process of the movement of water back and forward between land, oceans and the atmosphere is called the hydrological cycle.

◆ The hydrological cycle works as a closed system. There is a finite amount of water in the atmosphere, lithosphere and hydrosphere (land and water areas) which remains constant. The system is powered by energy from the sun.

◆ Within the system the amount of water within the various components (land, oceans and atmosphere) can and does vary, especially when the system is interrupted.

Key Point 2

Referring to a selected river basin you should be able to describe and explain the size and shape of the catchment area, the rainfall distribution and reliability, the surface features and the rock types.

Catchment area

◆ Depending on the basin selected for study, the size and shape of the catchment area can vary from a basin which is located completely within the boundaries of a single country or one which flows through several states or countries.

◆ Examples of the first type include the Amazon Basin in Brazil, the Hwang Ho basin in China, and the Indus basin in India.

◆ Examples of the second type include the Nile in North Africa, the Mississippi and Colorado basins in North America, and the Ganges basin in the Indian sub-continent.

◆ You should know the overall size of the selected basin in square kilometres and the area covered. If it flows through different states or countries you should be able to provide examples of these.

◆ You should be able to mention the main tributaries found in the basin.

Rainfall distribution and reliability

◆ You should have some general knowledge of the main features of the climate within your selected river basin. If the basin covers several climatic areas you should be able to refer to this.

◆ You should also be able to discuss rainfall distribution both throughout the year, referring to any periods where rainfall is excessive or unreliable. You should note the effect of this on river flow.

Surface features

◆ Rivers can flow through a variety of terrains. Basins such as the Nile, Ganges and Amazon will contain a variety of landscapes such as mountains and wide flat floodplains which terminate in wide deltas.

◆ These landscapes contain surface features which can interrupt and affect river flow such as waterfalls, rapids, gorges and areas of rainforest.

◆ Some of these surface features may offer opportunities for water control projects and the construction of multi purpose dams for water and electricity supply.

Rock type

◆ The variety of rock types will change considerably during the course of the river from source to mouth. The type of rock will have an impact on sites chosen for the construction of dams and reservoirs.

◆ You should have some general knowledge of the variety of rock types present within your selected basin and be able to comment on their impact on river flow and suitability for water control projects.

Case study
Nile Basin

Catchment area

◆ The source of the Nile is in East Africa at Lake Victoria.

◆ The main tributaries of the Nile are the White Nile and the Blue Nile.

◆ The river and its tributaries flow through several countries including Sudan and Egypt (Nile), Ethiopia (Blue Nile), Uganda and Sudan (White Nile).

◆ The Nile has an overall catchment area of 3 007 000 km².

◆ The length of the river is 6995 km.

◆ It is the world's second longest river, although its drainage basin is less than half that of the longest river, the Amazon.

◆ For the last 2700km of its course to the Mediterranean Sea the river flows through desert. The river terminates in the Nile Delta in the south-east corner of the Mediterranean Sea.

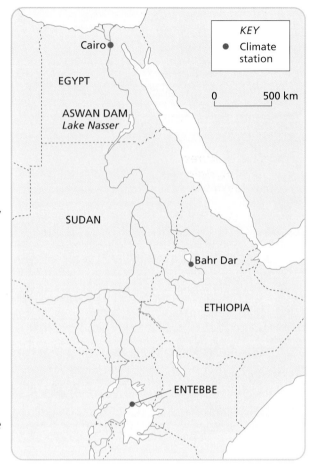

Figure 3.2.3 **The course of the Nile**

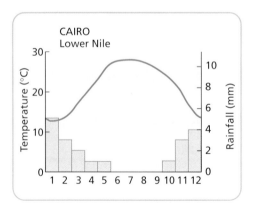

Figure 3.2.4 **Climate graph for Cairo**

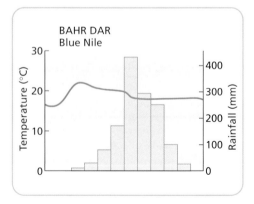

Figure 3.2.5 **Climate graph for Bahr Dar**

Rainfall distribution

◆ The source of the White Nile is in the highlands surrounding Lake Victoria in Uganda. The size of this lake means that the White Nile has a regular flow. This maintains water levels at a steady level. This helps compensate for the seasonality of the flow of the Blue Nile, which has its source in Ethiopia.

◆ There is a surge in the period from July to September. This is due to the pronounced seasonal regime of Blue Nile's catchment area in the Ethiopian Highlands (see Figure 3.2.5).

Surface features

◆ Both the White Nile and the Blue Nile flow through several different landscapes including mountains, plains and desert areas.

◆ In its lower course the river flows through a wide floodplain and terminates in the Nile Delta.

◆ In the Delta the river splits into several channels as it flows into the Sea.

Rock type

◆ The main rock type in its source region is volcanic.

◆ Throughout its course rock type changes from igneous to sedimentary.

◆ In the region of the Aswan dam, the major rock type is igneous. This provides a solid base for the construction of the dam and its reservoir.

Change to basin flow

◆ The flow of the river has been regulated since the Aswan dam was built and there are no extreme fluctuations which previously caused annual floods.

◆ Despite slight fluctuations, maximum discharge rarely exceeds 250 million cumecs (one cumec is one cubic metre per second).

◆ The dam provides irrigation for farmers in the Nile basin from Aswan to Alexandria on the coast.

Questions *and* Answers

?

Question 3.2.1

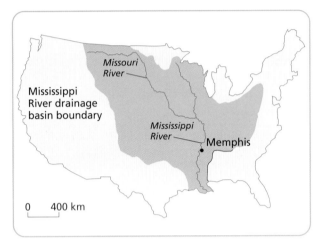

Figure Q3.2.1 a) The River Mississippi Drainage Basin; b) Annual Rainfall for Memphis, in the River Mississippi Drainage Basin

For the Mississippi River, or any other river you have studied, describe how the annual rainfall pattern affects the amount of water in the river.
(4 marks)

Intermediate 2

Answer

During the months of January and February rainfall is low and therefore the amount of water in the river may also be low (✓). This increases slightly during March and April when rainfall rises from about 25 mm to about 75 cm (✓). River water amounts may also rise (✓).

The rainfall rises to its maximum of just over 100 mm during June and July (✓) when the river will probably have its greatest amount of water (✓). The water level will probably decrease during August to December as the rainfall decreases during these months (✓).

Comments and marks obtained

This answer makes excellent use of the resource and quotes directly from the rainfall graph giving months when rainfall is high and low and makes reference to rainfall amounts. It also relates these to rising and falling levels of water in the Mississippi River with appropriate comments on increases and decreases. There are sufficient points for 6 marks. This answer therefore would gains the full **4 marks**.

Water control projects

Key Point 2

You should be able to describe and explain the social, economic and environmental effects of the water control project of your chosen basin and be able to assess the success of the scheme in terms of its social, political and environmental impact.

Social benefits
◆ Improved water quality leading to less disease and poor health and better food availability.
◆ More widespread availability of electricity.
◆ Greater population helped by increased food supply.
◆ More recreational opportunities.

Adverse social consequences
◆ Forced removal of people from valley sites.
◆ Increased incidence of waterborne diseases such as bilharzia (also known as schistisomiasis) and onchocerciasis, or river blindness.

Economic benefits
◆ Hydro-electric power, which is helping to create industrial development.
◆ More water for industry.
◆ Improved yields in farming.
◆ Improved navigation channels.

Adverse economic consequences
◆ Major expenditure on new schemes.
◆ Dependence on foreign finance resulting in increased debt.
◆ More money being needed for fertilisers (in agricultural schemes) and compensation for displaced people.

Environmental benefits
◆ Increased fresh water supply with consequent improvements in sanitation and health.
◆ Improvements in scenic value.
◆ Improvements in flood control.

Adverse environmental consequences
◆ Increased water pollution from new industries and new agricultural practices.
◆ Increased silting of reservoirs.
◆ Increased salinity rates further downstream.
◆ Possible flooding of historical sites.

Political benefits
◆ Improved political relations between adjoining countries and countries within the basin in general.

◆ Assistance from economically more developed countries to economically less developed countries.

◆ Improved communications, road, rail and water transport, assisting increased trade between countries within the basin.

◆ Increased use of the river for creation of shared resources such as electricity and irrigation.

Adverse political consequences

◆ Dependence on neighbouring countries upstream for water control.

◆ Complex legislation over appropriate water sharing by different states or countries.

◆ Reduced flow and increased salinity in some areas.

◆ Problems over allocation of costs and authority.

◆ Increased pollution across borders resulting in problems of allocating appropriate costs of cleaning the river.

Questions and Answers

Question 3.2.2

For any river water control project you have studied, describe its advantages and disadvantages either to non-agricultural users or to the environment. *(5 marks)*

Intermediate 2

Answer

Advantages of the Colorado water control project for non agricultural users include the provision of cheap electricity for domestic and industrial users (✓) from the HEP schemes which have been built (✓). The project also helps to reduce the threat of flooding (✓) and parts of the river are made more manageable for transport (✓). Disadvantages include increases in salinity levels which increases pollution levels (✓). Silt builds up behind reservoirs resulting in high replacement costs (✓) and many parts of the valley may be flooded by reservoirs causing loss of homes and scenery (✓).

Comments and marks obtained

This answer correctly identifies different types of non-agricultural users and gives several good examples of advantages such as cheap electricity, flood reduction and improved transport for 3 marks. Further marks are obtained for the disadvantages including silting, increased replacement costs, increased pollution levels and flooding due to reservoir construction. The answer is a high quality answer which obtains full marks.

Questions and Answers continued ➤

Questions *and* Answers *continued*

Question 3.2.3

Figure Q3.2.3 The Three Gorges Dam project

Study Figure Q3.2.3. Referring to the Three Gorges Dam or any other multi-purpose water project you have studied:

(i)   describe the economic advantages of the scheme.   *(3 marks)*

(ii) explain why many environmentalists are against multi-purpose schemes.
      (4 marks)

Intermediate 2 4b 2005

Answer

(i) *The economic advantages to the Three Gorges Dam would be that there is more water for industry. More water produces electricity to locals and to the industry (✓). Better navigation schemes, as the water flow was never accurate before.*

Comments and marks obtained

Unfortunately this candidate did not read the question properly. The question asked for 'economic advantages' such as creating power for industry and irrigation for agriculture which would result in greater industrial and agricultural output. This would increase the wealth of the country and its people. The candidate made only one valid statement about producing electricity to locals and industry which was just worth one mark. Reference to navigation might have gained a further mark had it been expanded by mentioning for example 'increased trade'. The answer gained only **1 mark out of a possible 3**.

Questions and Answers continued ➤

Questions and **Answers** continued

Answer

(ii) *Environmentalists are against these as they believe that the dams which are built scar the landscape (✓) and make it unnatural looking. They are also against it because it reduces water flow the front side of the dam disturbing wildlife and flooding behind the dam which disturbs wildlife (✓).*

Comments and marks obtained

The reference to scarring the landscape is valid and gains one mark. The comment on the reduction of water at the front is too vague to gain any further marks. However the reference to flooding behind the dam affecting wildlife is good enough for a second mark. The candidate could have gained further marks if he/she had referred to the loss of farmland, destruction of homes and other property and had elaborated on how floods affected wildlife (loss of habitat, drowning of animals). The answer therefore obtained only **2 marks out of a possible 4.**

Question 3.2.4

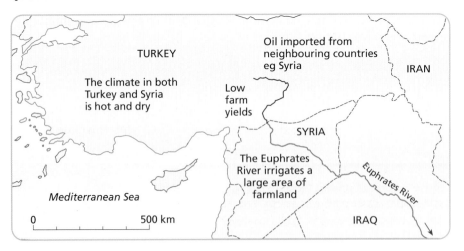

Figure Q3.2.4a The Euphrates River flows through Turkey, Syria and Iraq

Study Figures Q3.2.4a and b and their captions. Explain how the building of the Ataturk Dam in Turkey could cause political conflict with Syria. *(4 Marks)*

Intermediate 2 2005 4c

Questions and **Answers** continued ➤

Questions and Answers continued

Question 3.2.4 *continued*

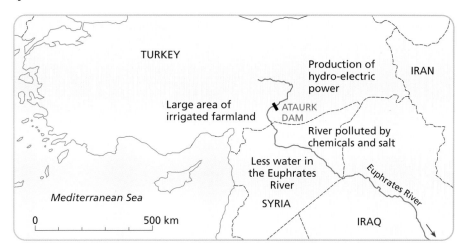

Figure Q3.2.4b Damming the Euphrates river in Turkey has consequences in Syria

Answer

The building of the Ataturk Dam would cause many political problems. The dam would cause pollution in the river in Syria and may cause problems to the farmers (✓) as they may pollute their animals. There may be arguments with who has to pay for the dam's expense. May cause problems with the flow of the river in Syria therefore may have consequences.

Comments and marks obtained

Unfortunately this answer made little use of the resource provided. A mark was gained for the reference to river pollution in Syria causing problems for farmers. However the candidate failed to expand on this by giving examples of the type of problems which this pollution could have caused. The brief mention of problems of river flow was insufficient for further marks. The candidate could have explained how interruption to river flow would have meant less water for irrigation in Syria. Mention could have been made of the effect on trade between the countries namely, oil imports from Syria. All of these points would have led to political tension between the countries. Consequently, the answer obtains **only 1 mark out of a possible 4.**

Glossary *River Basin Management*

Drainage: All surface water in a river system. (Do not confuse this with underground pipes used to drain water from bog or marshland!)

Environmental consequences: Net effects on the physical and human environments of changes to river basins through physical or human changes.

Infiltration: The process by which water from precipitation seeps into the soil and sub soil.

River basin: The water catchment area of a river. It includes the main river and its tributaries.

Salinity: Salt content in surface water (rivers, streams and lakes).

Seasonal fluctuation: Variations in weather patterns over the seasons of the year.

Silting: Sand and other material carried in solution in rivers. When deposited, silt can reduce river flow.

Tributary: A smaller river or stream which runs into a larger river

Water control project: Engineering schemes manage various aspects of a river basin such as river flow, reservoirs, silting, navigation and river discharge through the use of dams, canals, aquaducts and diversion schemes.

HOW TO PASS INTERMEDIATE 2 GEOGRAPHY

Section C
European Environmental Inequalities

This part of the syllabus is studied within the context of Europe as defined on Figure 3.3.1. When referring to case studies you have to refer to more than one country, at least one of which should be from mainland Europe. The topic areas are essentially concerned with environmental quality with specific reference to the quality of air, rivers, seas and coastal areas within Europe.

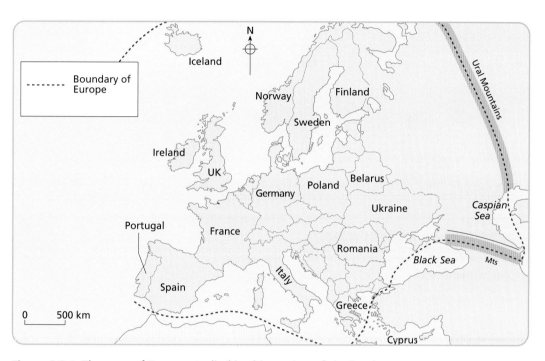

Figure 3.3.1 The area of Europe studied in this section of the book

Key Point 1

You should be able to describe patterns of environmental quality with reference to the quality of air.

Figure 3.3.2 illustrates patterns of environmental quality and is based on a map given in a recent examination question on this topic. Descriptions based on this map should refer to factors such as:

◆ differences in air quality throughout Europe;

◆ reference to specific items of data taken from the resource;

◆ reference and comparisons of specific countries within Europe, noting variations in the quality of air or pollution emissions.

133

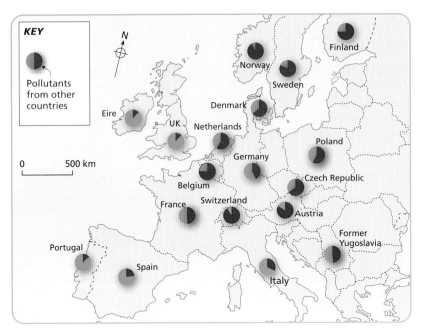

Figure 3.3.2 Patterns of environmental quality in Europe

Questions and Answers

Question 3.3.1

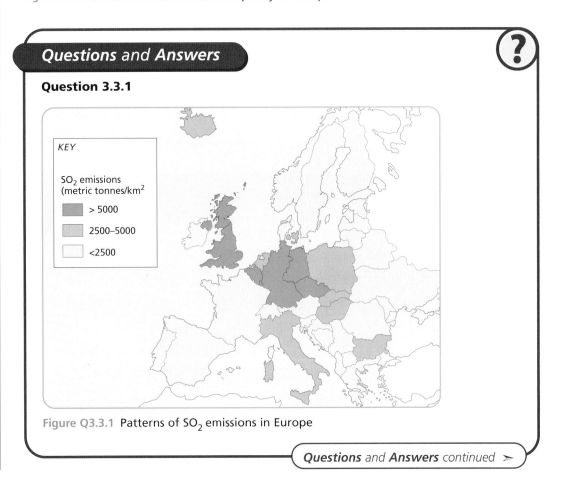

Figure Q3.3.1 Patterns of SO_2 emissions in Europe

Questions and Answers continued ➤

ENVIRONMENTAL INTERACTIONS

Questions and Answers *continued*

Study Figure Q3.3.1. Describe the pattern of sulphur dioxide (SO$_2$) emissions in Europe. *(3 marks)*

Answer

Sulphur dioxide is much more present in areas such as Britain and France where it is over 5000 tonnes/km^2 (✓). In Italy and Poland it is between 2500 and 5000 tonnes/km^2 (✓). In Spain, Portugal, Republic of Ireland it is under 2500 tonnes/km^2.

Comments and marks obtained

The answer gives three basic statements based on the resource provided. It refers to the level of emissions in three different areas on the map and in so doing notes differences between these areas. In effect it correctly describes the pattern of emissions using data from the resource. There is sufficient information for the answer to gain **3 marks out of 3.**

Key Point 2

You should be able to explain these patterns with reference to a range of physical, economic, social and political factors.

Population density

◆ Density of population can have an important influence on air quality. Areas with high population density are usually associated with the presence of large settlements and industrial complexes.

◆ Large settlements or urban areas can be the source of many forms of air pollutants including smoke from domestic chimneys, pollutants from road traffic such as cars, buses and commercial vehicles in the form of exhaust fumes, public waste incinerators and pollution from industrial chimneys, giving off gases such as sulphur dioxide and carbon monoxide.

◆ Waste tips may also be a source of air pollution from gases such as CFCs which can escape from discarded domestic appliances such as refrigerators.

Greenhouse effect and global warming

◆ When carbon dioxide is released into the atmosphere through the burning of fossil fuels such as coal, oil and gas, this contributes to the greenhouse effect (see Figure 3.3.3).

◆ Carbon dioxide allows energy from the Sun through to reach the surface of the Earth, but absorbs heat radiated from the Earth to the atmosphere.

◆ As this gas builds up in the atmosphere, the air temperature increases as does the temperature of the Earth.

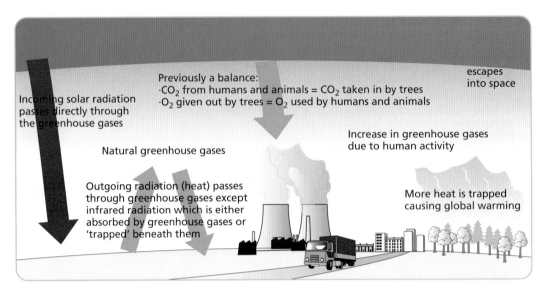

Figure 3.3.3 The Greenhouse Effect

- During the last 100 years, the temperature of the Earth has increased by 0.5 degrees. This is known as *global warming*. Scientists predict that if this continues, it could lead to major changes in sea levels due to melting ice caps, and changes in ocean currents such as the North Atlantic Drift which could in turn lead to another Ice Age.

- Sea level changes would be disastrous to low lying coastal areas, which often contain large populations.

Transport links

- Major transport links in countries are a source of air pollution due to exhaust fumes giving off toxic gases such as nitrogen oxides, carbon monoxide and sulphur dioxide.

- As the volume of traffic increases with the construction of new roads and motorways, levels of pollution are increasing throughout Europe.

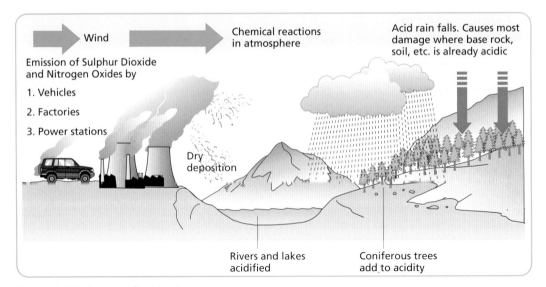

Figure 3.3.4 Causes of acid rain

Physical and climatic environment

◆ Changes to the physical and climatic environments, such as deforestation, and tourist developments in mountain areas have exposed new areas to air pollution.

◆ Destroying forests for commercial purposes reduces the amount of carbon dioxide taken from the atmosphere and contributes to the greenhouse effect and global warming.

◆ Few areas in Europe are pollution free.

Economic activities

◆ Economic activities such as farming can lead to air pollution from sources such as pesticides, insecticides, chemical fertilisers and methane gas given off from livestock.

◆ Overall pollution from farming, industries and activities such as tourist developments is continually damaging the quality of air throughout the European continent.

Living standards

◆ Increasing living standards causes increased demand for fossil fuels for transport purposes (such as increased air travel for new holiday venues).

◆ There is increased demand for land for development in previously unused areas for holiday and other purposes.

Public attitudes

◆ Many governments recognise the need for policies which encourage the public to respect and protect the environment, but these efforts are not always successful. It is often difficult to enforce legislation due to lack of manpower and legal costs.

◆ Raising public awareness of the consequences of damage to the environment is perhaps the best way to tackle this issue.

Rivers

Key Point 3

You should be able to describe patterns of environmental quality in relation to rivers within Europe.

Key Point 4

You should also be able to explain these patterns in relation to factors similar to those responsible for air pollution such as those shown on Figure 3.3.5

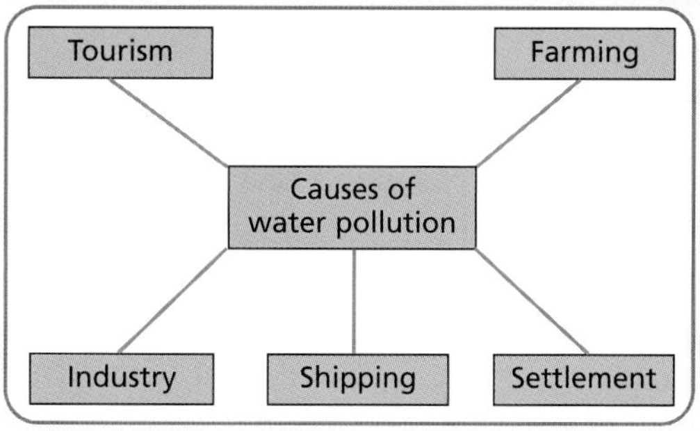

Tourism — Farming

Causes of water pollution

Industry — Shipping — Settlement

Figure 3.3.5 **Causes of river pollution**

Population density

◆ Many cities and towns are located on the banks of major rivers within Europe. These settlements have regularly discharged sewage and industrial waste into rivers, causing the rivers to become so polluted that plant and animal life in the rivers has become extinct.

◆ When a river becomes polluted, oxygen levels fall to the extent that the life within and around the river is unable to survive.

◆ Untreated sewage carries bacteria which can spread infection and is foul smelling.

◆ Many urban areas created landfill sites which were filled with rubbish which often included toxic chemicals. Rainwater draining through these landfill sites often carried these chemicals away and eventually this drained into rivers, adding to pollution levels.

Transport links

◆ River valleys have always been used as main transport links throughout Europe. Most forms of transport (road, rail and water) have used the valley's flat land for communication.

◆ In the past when trains ran on coal or oil, large quantities of pollution were given off which eventually dissolved in the rivers. Similarly fumes from motor vehicles contribute to river pollution.

◆ Whenever rivers were used as a means of transport for barges and ships, discharge from engines and leakages from cargo holds often found their way into the water, thereby increasing pollution levels.

◆ Major rivers such as the Rhine, Thames, Clyde and Seine which had major port developments gradually became very badly polluted from the traffic both on and around the rivers.

Economic activities

◆ River valleys offered a wide variety of opportunity for human development such as settlement, industry, communications and agriculture. The wide, flat floodplains were ideal sites for these developments. Consequently both the valleys and the atmosphere above often became contaminated due to exposure from all of these activities.

◆ Towns, cities and ports often used rivers as a means of disposing of waste materials with little consideration to the damage caused to the environment and the consequences for future generations.

◆ Similarly, other economic activities such as farming and tourism also contributed to river pollution. For example, chemicals used by farmers to increase crop production often drained into rivers.

◆ River valleys have long been popular with tourists in providing sites for camping and caravans. The rivers were an obvious source for waste disposal from these sites.

Living standards

◆ The affluent society has often been described as the 'effluent society'.

◆ Chemical works, power plants, sewage works, large iron and steel works, shipyards, dockyards, petrochemical factories etc., have all been attracted to river valley sites.

◆ All have at some time contributed to the amount of pollution in rivers either through using water from rivers in industrial processes and then returning polluted water to the river or by discharging waste and having leakages which find their way into the water.

◆ Whilst all of these activities were necessary to develop the wealth of these communities, the environmental cost in terms of destroying the rivers and their valleys was often forgotten or overlooked in the past.

Questions *and* Answers

Question 3.3.2

With reference to two rivers/sea and coastal areas or two mountain areas you have studied, explain the differences in environmental quality between them. *(6 marks)*

Intermediate 2

Answer

Two coastal areas I have studied are the north west of Scotland and the Costa Brava in Spain. Unlike the Costa Brava the environment in Scotland is quiet and peaceful since there are very few people living there (✓) and there are few industries to pollute the coast or spoil the scenery (✓). The area is more remote than the Costa Brava and therefore does not attract tourists who might pollute beaches with litter (✓✓).

The Costa Brava is a popular, busy tourist resort and suffers from pollution from sewage, beach litter and traffic congestion (✓). Farmers use fertilisers to grow more crops for tourists and some of this runs off into the Mediterranean Sea (✓). Many large ships dock in ports along the Costa Brava creating even more pollution and therefore spoiling the quality of the environment (✓).

Comments and marks obtained

This answer begins well by immediately comparing the two areas. References to few people contrasts with later statement on 'busy' Costa Brava, as does the reference to few industries in north west Scotland with 'farmers polluting the Mediterranean' and 'large ships creating pollution'. There are sufficient references to the differences in quality of environment between the selected coastal areas to merit **full 6 marks**.

HOW TO PASS INTERMEDIATE 2 GEOGRAPHY

Seas and Coasts

Key Point 5

You should be able to describe patterns of environmental quality throughout Europe's major seas and coastal areas, as for example those shown in Figure 3.3.6.

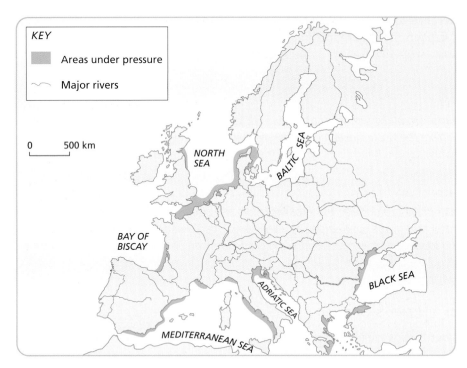

KEY

- Areas under pressure
- Major rivers

0   500 km

NORTH SEA

BALTIC SEA

BAY OF BISCAY

ADRIATIC SEA

BLACK SEA

MEDITERRANEAN SEA

Figure 3.3.6 **Patterns of environmental quality in seas and coastal areas around Europe**

In examination questions, your answers should refer to areas where the level of environmental quality is low, medium or high. Refer to specific data given in the question to support your statements. Refer to specific countries if these are named on any map provided as a resource.

If specific areas are not given, you can name these areas if you know them or give general locations such as 'Southern Europe', 'North East Europe' etc. Refer to general patterns within the map provided.

You should also describe patterns within specific areas. Draw comparisons between countries using data provided. Refer to areas of good or poor quality environments within specific regions or countries such as Western Europe, Northern Europe or Eastern Europe. If there are any figures given, make references to these.

Key Point 6

You should be able to explain patterns of environmental quality in relation to the factors discussed above in connection with air quality and rivers. In your explanation refer to factors such as population density, economic activity and living standards.

Population density

◆ Coastal areas are highly populated since they offer great opportunities for urban and industrial development.

◆ Seas are used for transport and for industries such as fishing. Major ports such as London, Marseilles, Rotterdam and Lisbon have grown, attracting populations in huge numbers by the offer of employment in a wide range of port related industry. As with river valleys, these settlements have become a major source of pollution due to chemical, sewage, industrial and shipping discharges.

◆ Many sea areas have become badly polluted due to the discharge of raw sewage and industrial waste such as mercury, lead and aluminium deposits, from pipes leading from the ports and cities out into the sea and factories built along coastlines. Often visual evidence of this can be seen along many miles of beaches in the form of debris and sewage which has been washed up from the sea.

◆ As the volume of shipping increased due to industrial development, so also did the volume of pollution in the European seas.

◆ Waste and fuel discharges from large ships such as tankers have greatly affected seas and marine life within them. Accidents involving oil tankers have caused immense damage to seas and to large areas of coastline.

◆ Fish and other wildlife such as birds have been killed in vast numbers due to oil spillages. This pollution can affect the food chain within seas.

◆ Pollutants absorbed by fish which are subsequently eaten by humans can lead to severe illnesses including cancers.

◆ Oil sinking to the sea bed affects plant life which is necessary to the survival of fish and other animals.

Economic activities

◆ Many seas around Europe are relatively shallow and are unable to cope with the ever-increasing amounts of pollutants.

◆ Developments such as drilling for oil in the North Sea have increased the pressure in these waters.

◆ The Mediterranean Sea revitalises itself through the flow of water from the Atlantic Ocean through the Straits of Gibraltar. Anything which affects this inward and outward flow can seriously threaten the future of this sea area. Changes to salinity levels due to pollution is one such threat.

ENVIRONMENTAL INTERACTIONS

◆ The Mediterranean is a very popular tourist area with millions of tourists visiting the many countries with a Mediterranean coastline. This increases pressure in many ways such as increased sewage disposal, discharge from local farming which is anxious to increase crop output to meet food demands from the tourist industry and increases in litter on tourist beaches in the area.

◆ Industrialists have in the past used the seas as dumping grounds for industrial waste including radioactive waste. This particular waste could take up to thousands of years before it becomes harmless.

Living standards and attitudes

◆ As living standards improved throughout Europe, this has led to a massive increase in the tourist industry.

◆ Coastal areas suffer more than most from the demands of this industry. Regardless of climate, vast numbers of tourists are attracted to coastal resorts throughout Europe.

◆ Increased pressure from this industry on sewage disposal, water extraction from rivers, waste disposal, increased farming output and increased pressure on all forms of travel and communications has greatly impacted on the quality of environment in sea and coastal areas.

Key Point 7

You should be able to describe and explain differences in environmental quality between two rivers (one of which should be from mainland Europe).

In class you may have studied various river basins as case studies. Two river valleys which you could have studied may have included the River Tees in North East England and the River Rhine in mainland Europe. Figure 3.3.7 shows the locations of these rivers.

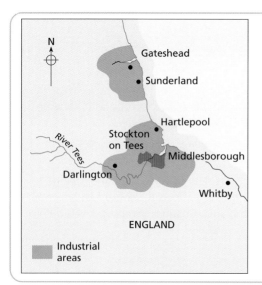

Figure 3.3.7 Location of River Tees (UK) and River Rhine (Germany)

Key Point 8

When comparing these rivers in terms of environmental quality you could refer to the following:

◆ the general characteristics of the valleys such as the length of the river, number of countries through which the river flows, the number and type of settlements located along its course;

◆ the main types of pressures and the source of these pressures;

◆ efforts and strategies adopted to manage, improve and maintain environmental quality on the rivers;

◆ the impact of these efforts and comments on the relative effectiveness of local, national and international policies to manage, improve and maintain environmental quality.

For the River Tees you might refer to:

◆ the quality of the river and its valley in terms of levels of pollution and cleanliness;

◆ changes in the quality of the river for fish and other wildlife;

◆ local wildlife protection areas situated on the river;

◆ land use along the course of the river, with particular reference to industry and housing.

For the River Rhine you might comment on any differences which exist between the Tees and the Rhine such as:

◆ differences in pollution levels between the two rivers;

◆ differences in the immediate human landscape in terms of industry and housing;

◆ differences in the use of the rivers as major waterways for shipping, particularly oil tankers;

◆ differences in population density in the river valleys;

◆ differences in the success of efforts to improve on environmental quality between the two rivers;

◆ differences in the local and national agencies involved in quality improvement efforts.

Key Point 9

You should know about the differences in environmental quality between either any two sea areas and coastal areas, or two mountain areas in Europe. Figure 3.3.8 shows the locations of mountain areas under threat.

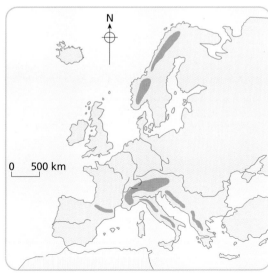

Figure 3.3.8 European mountain areas under environmental pressure

If in class you are asked to study two sea or coastal areas, your studies might include for example the North Sea coast and the Mediterranean Sea coasts. Similarly, if you study two mountain areas, these might include the Alps and the Lake District in North West England.

Note that your answer can refer specifically to *either* sea areas *or* coastal areas. Your discussion of these areas would refer to general characteristics such as location, main land uses, general environmental pressures and the causes of these pressures.

You would also wish to describe and assess the effectiveness of local national and international policies designed to manage, improve and maintain environmental quality within these areas.

Key Point 10

For your case studies of two sea areas such as the North Sea and Mediterranean you might include reference to:

◆ the number of countries and examples with access to the seas;

◆ pollution from industry, agriculture and tourism;

◆ economic impacts.

The North Sea

Countries with coastlines on the North Sea include the UK, Iceland, Norway, Denmark and the Netherlands (see Figure 3.3.6). The main environmental pressures affecting the North Sea are related to:

◆ oil exploration, over-fishing, the dumping of sewage;

◆ industrial waste and chemicals from agriculture;

◆ waste and accidents from international shipping;

◆ pollution from tourist activities.

The Mediterranean Sea

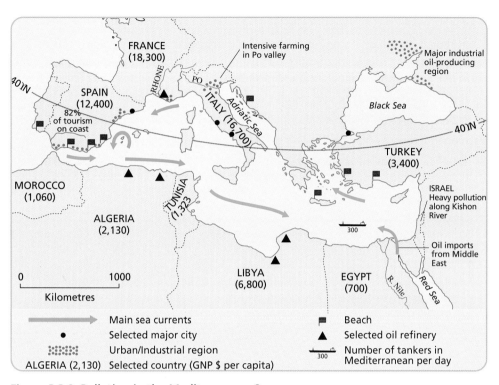

Figure 3.3.9 Pollution in the Mediterranean Sea

The Mediterranean Sea has 13 countries with a coastline on this sea area. European countries with a Mediterranean coastline include France, Spain, Italy and Greece. The main environmental pressures affecting the Mediterranean Sea are similar to those experienced in the North Sea. The main differences are due to the increased pressures caused by far greater levels of tourist activities in the Mediterranean.

◆ Unlike the North Sea, millions of tourists visit the resorts of European Mediterranean countries and this places great demands on other areas of the economy.

◆ These include demands on activities such as farming, water supply, industry, communications and urban waste disposal.

◆ Consequently the Mediterranean has had vast quantities of pollutants dumped through shipping, farm chemicals, sewage and industrial waste and beaches strewn with litter and other harmful materials left by tourists.

Management of environmental quality in North Sea

Efforts to manage, improve and maintain environmental quality in the North Sea have included:

◆ monitoring of sewage disposal and industrial waste by government agencies;

◆ plans to clean beaches;

◆ imposition of fishing quotas under national and European laws;

◆ a variety of protection measures implemented at local level;

◆ protests led by environmental groups such as Greenpeace.

Management of environmental quality in Mediterranean Sea

The Mediterranean countries have signed up to the 'Blue Plan' policy. Measures in this plan include:

◆ awarding 'Blue flags' to countries which have managed to obtain a high standard of water quality through improvements in the provision of cleaner beaches and reductions in levels of industrial and agricultural pollution;

◆ fining countries or regions where pollution levels are unacceptable;

◆ providing agencies which monitor pollution levels and standards of cleanliness;

◆ monitoring waste from ships using the sea, especially oil tankers;

◆ adhering to guidelines on reducing levels of sewage and waste disposal issued by the European Union have helped to greatly improve environmental quality in this sea area.

Key Point 11

You should be able to comment on the effectiveness of local, national or international policies and strategies to manage and improve environmental quality.

Sea areas

◆ For both sea areas, environmental pressure groups have been successful in improving public awareness of the problem. Groups such as Greenpeace make great efforts to highlight environmental problems caused by industry and tourism.

◆ For example, in October 2004 Greenpeace organised a demonstration in the English Channel to prevent ships loaded with plutonium ores from America from docking in Cherbourg, France.

◆ Beaches which remain polluted are named and shamed with the effect that they are less likely to remain attractive to tourists. Water and beach quality are assessed regularly and their results are given high profile.

◆ These and similar measures are gradually having a significant impact in improving environmental conditions, fish stocks and beaches within these sea and coastal areas.

Two mountain areas

Pressures in mountain areas are also very much related to the increase in tourism and associated activities.

◆ Large numbers of tourists travelling from lowland to mountain areas cause increased traffic. The delivery of services required by tourists also increases traffic.

◆ Scarce land resources are needed for hotels and other accommodation for tourists.

◆ Hill walkers and mountain bikers can cause increased erosion of the physical landscape.

There are major differences between mountain areas in Europe. For example, the Highlands of Scotland are less spoilt than areas within the Alps. Part of this is due to the fact that the Scottish Highlands are relatively more remote than the Alps. There are fewer winter tourists resorts in the Highlands due to the unreliability of snowfalls compared to the Alps. The skiing season is more predictable and longer in the Alps. Consequently the Alps attract thousands more tourists than the Scottish Highlands.

The Alps by comparison suffers a great deal more pollution from traffic, tourists and visual pollution on mountain sides from tourist facilities such as hotels and ski runs.

There is also land use conflict between environmentally friendly land uses such as farming and the use of remote areas for activities such as military training.

Key Point 12

Possible case studies of mountain areas include the Lake District in England (Figure 3.3.10) and the Alps in Europe (Figure 3.3.11).

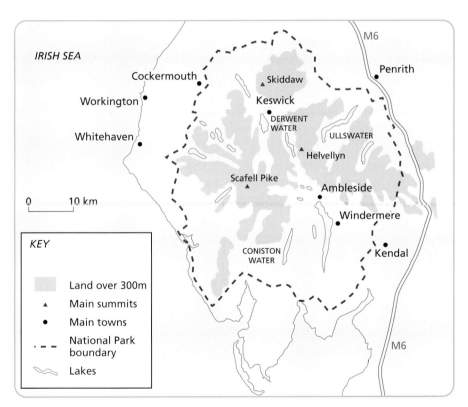

Figure 3.3.10 **The Lake District**

The Lake District

◆ There is a wide variety of land uses within the Lake District. These are often in conflict with the environment and with each other. These land uses include farming, settlement, transport, water supply, power supply, industry, military training and leisure and tourism.

◆ In order to protect and maintain the environment, this area has been designated a National Park and is under the control and protection of the National Park Authority.

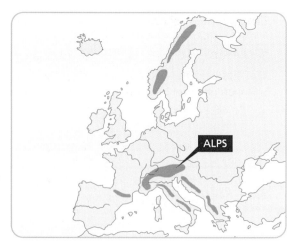

Figure 3.3.11 **The Alps**

The Alps

◆ Within the Alps, pressures are related to changes in farming, an increase in roads and pollution from traffic.

◆ Other pressures include deforestation to make room for tourist developments, visual and noise pollution from the construction and provision of facilities to cater for a growing tourist industry.

◆ Villages which were once remote and relatively free from environmental pressure are now being taken over by the tourist industry and have effectively become mountain tourist resorts. Although this may bring wealth and economic prosperity to these areas, the price is a clear reduction in the quality of the environment.

◆ Unlike the Lake District, these areas are not under the protection of organisations such as National Park Authorities. For protection the area depends on the efforts of national and local government authorities and environmental pressure groups to protect and sustain the natural environment.

Main differences between the environmental quality of the Lake District and the Alps

The standard of environmental quality in the Lake District is better than in the Alps. Levels of pollution are lower. This includes different forms of pollution such as visual, water and land pollution.

Reasons for differences

◆ The Lake District is under the protection of a National Park Authority. This Authority has the legal right to limit development, offer advice on the protection of the environment through education centres, and provides wardens and rangers to monitor instances of abuse and misuse.

◆ In the Alps, this kind of Authority does not exist for the whole area.

◆ The scale of operations is much greater because the Alps cover several countries such as France, Italy, Switzerland and Austria. International cooperation is essential if standards of protection and conservation are to be maintained.

Effectiveness of national and international policies

◆ The effectiveness of implementing management strategies is hampered in the Alps because of the size and the degree of international cooperation required.

◆ Those countries which are part of the European Union, France, Italy and Austria must implement and adhere to European laws designed to protect the environment, as does Britain. Switzerland is not a member of the European Union, so it does not have to adhere to the same laws.

◆ It is much easier to monitor and implement laws on a smaller scale and within the same country.

◆ New legislation on environmental control issued through the EU is having an impact on ensuring a general improvement in standards in Britain and especially in environmentally sensitive areas such as the Alps.

Glossary   *European Environmental Interactions*

Acid rain: Rain which contains acid in solution. The acid is formed usually from gases given off from sources such as exhaust fumes and coal-fired power stations which give out sulphur dioxide and nitrogen oxide.

Air pollution: Air becomes polluted when chemical substances are released from sources such as motor vehicles, industries, power stations, burning forests and domestic chimneys. These chemicals are absorbed into the air, sometimes in solution with rain. This can cause problems for vegetation, river and lake wildlife and humans.

Blue Plan: A plan designed to improve the quality of water and coastal areas around the Mediterranean. It imposes a wide variety of measures to prevent pollution from various sources such as industry, farming, tourism and domestic sewage disposal. Governments in these areas took legal action against polluters and made efforts to control and clean contaminated areas. The Plan has met with limited success.

Contamination: When foreign substances such as industrial and chemical waste products are discharged into air, water and land areas this leads to impurities being absorbed which seriously affects the quality of the original hosts (air, water and land sites). The host areas are said to have become 'contaminated' by these substances.

Deforestation: Large scale removal of forest areas throughout the world ranging from tropical rainforests to temperate forests. Trees are removed for many reasons, particularly commercial purposes causing great damage to the natural environment and habitats of animals and humans.

Environment: The totality of the surrounding area, which may consist of the physical, or natural environment, the human environment or a combination of both physical and human environments.

Environmental quality: The extent to which any environment may be affected by different forms of pollution such as air or water pollutants or visual pollution such as large inappropriate buildings.

Food chain: The life system in which various forms of life provide food for each other, beginning at one point and finishing at another. For example, water weeds are eaten by small fish which are eaten by larger fish which are eaten by humans.

Fossil fuels: Organic materials such as coal, oil and gas which derive from the remains of long dead animals and plants. Fossil fuels are a major source of fuel for power stations.

Global warming: As a result of the greenhouse effect, temperatures are rising throughout the world giving rise to climate changes and rises in sea levels due to melting ice caps.

Greenhouse effect: The warming effect due to the presence of gases like carbon dioxide and methane in the atmosphere. Increased levels of these gases as a result of burning fossil fuels (coal, oil and gas) and as a result of deforestation is leading to global warming and climate change.

Industrial waste: Material dumped and discharged from large factories into rivers, seas and selected land areas. It can consist of anything from fossil fuel remains to chemical waste from industrial processes.

Glossary *European Environmental Interactions continued*

Insecticides: Chemicals designed to kill insects which may destroy crops and other vegetation. The insecticide may be sprayed over large areas from planes. Some of these chemicals may find their way through absorption of crops or consumption by animals into the human food chain, often causing disease in the process.

Landfill sites: Areas set aside for dumping and burying industrial and domestic waste. They contain large pits which are covered over with soil once they are filled with waste. If they are not built correctly, the soil underneath the waste may break down into chemical solution which can penetrate the subsoils and can make its way into the water table.

Over-fishing: Depletion of fish stocks in oceans and seas at a rate greater than the rate at which fish can reproduce. This leads to an overall reduction in the number of fish which can be caught. Over-fishing often involves catching young fish before they have a chance to breed. Very large fishing boats called 'factory ships' can catch far greater numbers of fish than typical fishing boats and are a major cause of over-fishing.

Pesticides: Chemicals used in farming to kill various species of insects which can destroy crops. These chemicals can eventually get into the food chain with very harmful effects for animals and humans.

Pollutants: Material released into the environment which ultimately causes damage to land, water areas and the atmosphere.

Pollution: Contamination of the natural environment with harmful substances as a result of human activities.

River pollution: The discharge of industrial waste and chemical substances into rivers. The purity of the water in rivers can be badly affected by river pollution.

Sewage: Waste material from domestic and industrial premises, which can badly damage sea areas and their coastlines if it is not treated properly before being discharged.

Visual pollution: Unsightly buildings, power stations, and industrial premises can destroy the appearance of an area, particularly in countryside areas.

ENVIRONMENTAL INTERACTIONS

Section D
Health and Development

The regional context for this topic is global. This means that you can quote examples from any country in the world.

Key Idea 1

You show know what is meant by *development* and recognise a selection of development indicators and be able to use them to classify countries as economically more or less developed.

Development refers to the level of social and economic status of a country. In other words whether a country is socially and economically 'rich' or 'poor'. Any improvement in the standard of living of people is called development.

Key Point 1

You should be able to identify economic and social indicators of development and show how these can illustrate different levels of development of countries.

Economic indicators

◆ Gross National Product (GNP) or Gross Domestic Product (GDP) statistics.

◆ Data relating to average income per capita.

◆ Data relating to the relative percentage of the workforce employed in industry and agriculture.

◆ Figures showing steel production (tonnes/capita).

◆ Average electricity consumption (kilowatts/capita).

◆ Trade patterns in terms of import and export figures.

◆ Trade balances in terms of surplus or deficits.

Social indicators

◆ Birth rates/death rates/infant mortality rates/life expectancy rates.

◆ Population structure in terms of the distribution of age and sex of the population.

◆ Average calorie intake per capita.

◆ The average number of people per doctor.

◆ Literacy rates as an indication of the level of education.

Note that decisions on levels of development should be based on a number of indicators used together rather on individual indicators such as GNP. The use of single indicators can be misleading since the data is based on averages and do not reveal the whole situation.

For example, average per capita income or GDP per capita for Saudi Arabia may seem high and suggest a high level of development but income distribution is very uneven varying from extremely high to very low. Similarly GNP may indicate a high level but be based on a single commodity such as oil. Taken together to produce combined indices such the 'physical quality of life' index or the Human Development Index (HDI), this gives a much more accurate picture of the stage of development of any given country.

Combined indicators

Since individual indicators are not always reliable for defining the level of development of a country, it is generally accepted that a combination of indicators gives a more accurate picture. Combined indicators include:

◆ The Human Development Indicator (HDI). This is a combination of indicators such as life expectancy, literacy, cost of living, school enrolment and GNP per person. The index ranges from 0 to 1.

◆ Physical Quality of Life Index (PQLI). This combines life expectancy, infant mortality and literacy to produce an index from 0 to 1. Figure 3.4.1 shows levels of development of countries throughout the world based on the Physical Quality of Life Index.

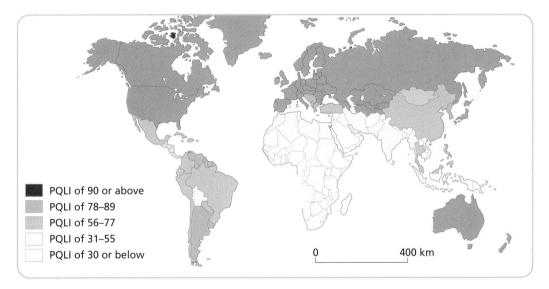

PQLI of 90 or above
PQLI of 78–89
PQLI of 56–77
PQLI of 31–55
PQLI of 30 or below

0   400 km

Figure 3.4.1 Levels of development based on the Physical Quality of Life Index

HOW TO PASS INTERMEDIATE 2 GEOGRAPHY

Questions and Answers

Question 3.4.1

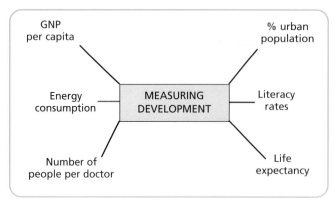

Figure Q3.4.1 Measuring Development

Study Figure Q3.4.1 above. Select one economic and one social indicator from the diagram. For each, explain how it can show the level of development. *(4 Marks)*

Answer

One economic indicator on the diagram is GNP per capita. This is the average amount of production per person in a country (✓). This is a good way of showing the wealth of the country overall (✓). A social indicator is literacy rates which shows the percentage or number of people per thousand who can read and write. This is a good indicator since high literacy rates would mean that a large number of people have good levels of education (✓). It also shows that the country has plenty of schools and teachers, which shows a high level of development (✓).

Comments and marks obtained

This is a high quality answer which correctly identifies an economic and a social indicator from the diagram, although there are no marks for identifying these. However, the candidate explains what both indicators show in general about levels of development and therefore obtains two marks for this. A further two marks are gained by elaborating in what ways these indicators can show both the overall wealth of a country and how high levels of education are achieved. The answer gains a full 4 marks out of 4.

Key Point 2

You should be able to describe and explain differences in levels of development and explain the limits of some indicators such as GNP in accurately reflecting different levels of standard of living within any one country.

Different levels of development

◆ Countries such as Saudi Arabia, United Arab Emirates and Brunei have prospered due to oil and gas reserves.

◆ Singapore, South Korea and Taiwan have encouraged the development of industry and commerce due to their entrepreneurial skills and have prospered.

◆ Countries such as Ethiopia or Chad lack natural resources and experience recurring drought leading to famine.

◆ Some countries such as Bangladesh suffer natural disasters such as floods and cyclones.

◆ Political instability, problems of rapid population growth and civil disorder also affect economic growth in many developing countries.

Accuracy of development indicators
As noted earlier, the use of one individual factor could be misleading especially if that factor is based on average figures, such as income per capita or GNP. This does not reveal the possible wide variations existing within a country with some people very wealthy whilst others live at subsistence level.

◆ For example, in some Middle Eastern oil-producing countries, GNP might appear to rank alongside those of highly developed countries but this wealth is not evenly spread throughout the population.

◆ Caution must also be used when looking at some indicators relating to social and economic development levels in for example Brazil or India.

◆ Therefore using a combination of indicators to produce a quality of life index is the best method of assessing levels of development in any given country.

Key Idea 2

You should know the causes of different levels of development.

Key Point 3

You should know the main physical factors affecting the level of development of a country. These include climate, relief, resources, environment and natural disasters.

Climate
◆ Very wet climates can cause natural disasters such as floods in monsoon climates in India and Bangladesh.
◆ Very dry climates may lead to droughts. In countries like Ethiopia and Sudan, crops die due to a lack of rainfall and food supplies become scarce often leading to famine.
◆ Countries with adequate rainfall and moderate temperatures are highly suitable for various economic enterprises such as farming, industry and tourism.

Relief
◆ Areas with highland and steep slopes are often difficult to develop.
◆ Industrial and urban development requires flat land for building. Similarly transport routes may be affected.
◆ The type of farming in highland areas may be restricted to livestock farming such as hill sheep or herding which are less profitable than commercial arable and dairy farms.
◆ Areas where the relief is flat have a great advantage for economic and human development.

Resources
◆ Resources are often related to wealth and countries such as the USA with an abundance of resources are the most developed.
◆ Some parts of the world are rich in a variety of resources such as mineral resources, vegetation, climate and people. In such areas industry and urban areas develop and the areas become prosperous.
◆ Other parts of the world lack resources and are much less developed with poorer economies.
◆ Other countries which have certain resources such as oil may be wealthy but unfortunately the wealth is not evenly distributed among the general population.

Environment
◆ Countries with heavily polluted environments may indicate a high level of development.
◆ The high levels are often due to the presence of industries which pollute both the landscape and the atmosphere.

- Countries which have a large part of their population living in cities often have the worst living environments.
- Cities may be polluted due to traffic congestion, urban decay and dereliction.
- Many ELDC cities have major environmental problems resulting from the existence of shanty towns, squatter areas or areas where housing is located in dangerous areas such as the lowest parts of floodplains.

Natural disasters

- Some parts of the world suffer from various types of natural disasters such as floods, tropical storms, earthquakes, volcanic eruptions and droughts.
- Whenever these occur, they can disrupt the infrastructure of the country by destroying communications, farming, towns and cities and industrial areas.
- Often these disasters are common to the poorer, less developed countries of the world such as the Caribbean islands, parts of Africa, South East Asia and India.
- Despite great efforts by international agencies to help such areas develop, many years of effort can be destroyed by a single natural disaster.

Key Point 4

You should know the main human factors which affect the level of development of a country. These include demographic changes, urbanisation, industrialisation, trade and technology. These factors are summarised in Figure 3.4.2.

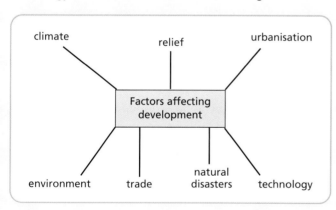

Figure 3.4.2 Factors affecting development

Demographic changes

- Changes in population numbers (increases or decreases) can seriously affect levels of development.
- If a country is overpopulated, this places great demand on the resources of the government to provide appropriate health care, education, employment, housing and public services.
- If a country is underpopulated, its economic development can be affected by a lack of people to work in industry or farming and to support those who are economically dependent.

◆ An imbalance in the population structure can result in either too many young or old people. Both situations place great strain on a country's economic development.

Urbanisation

◆ Urbanisation is the process whereby a large part of the population lives within cities and towns as opposed to rural areas.

◆ The percentage of urban population is increasing in both EMDCs and ELDCs.

◆ Population is attracted to urban areas for a variety of reasons, including employment, better housing, health and educational opportunity. Wherever these opportunities are limited, the level of development can deteriorate.

◆ Instances of homelessness, high unemployment, disease, poverty, high birth, death and infant mortality rates are a common feature of countries with lower levels of development.

Industrialisation

◆ Countries which have a higher percentage of the population employed in industry, particularly manufacturing and service sectors, usually have a high level of social and economic development.

◆ Countries in which a large proportion of the working population are employed in agriculture are usually those where the level of economic and social development is low.

◆ The distribution of employment is usually accepted as a very good indicator of development status.

Trade

◆ Those countries which trade mainly in manufactured products as opposed to primary products are normally regarded as highly developed.

◆ Those which trade mainly in primary products such as agricultural produce, timber and raw materials are normally regarded as less developed.

◆ Many less developed countries have a trade imbalance in that they import more goods than they export.

◆ Consequently such countries often rely on aid programmes to supplement their lack of income from trade.

Technology

◆ Levels of technology are a clear indicator of development. Highly industrialised countries use sophisticated machinery at the highest level of technology.

◆ Farming in more developed countries is much more mechanised. Less developed countries rely much more on human labour in farming than on mechanisation.

◆ This consequently impacts on farm yields and general food supply, with more developed countries having much higher yields and a reliable supply of food than less developed countries.

Questions *and* Answers

Question 3.4.2

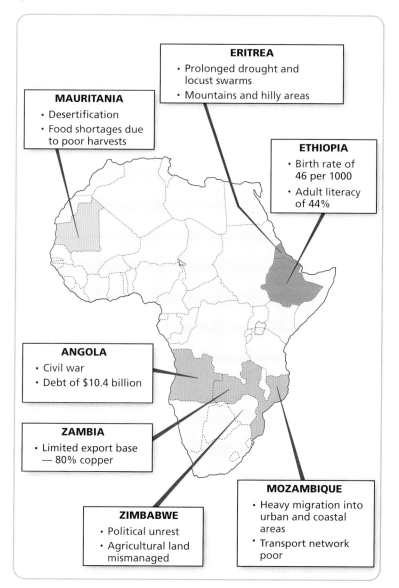

ERITREA
• Prolonged drought and locust swarms
• Mountains and hilly areas

MAURITANIA
• Desertification
• Food shortages due to poor harvests

ETHIOPIA
• Birth rate of 46 per 1000
• Adult literacy of 44%

ANGOLA
• Civil war
• Debt of $10.4 billion

ZAMBIA
• Limited export base — 80% copper

MOZAMBIQUE
• Heavy migration into urban and coastal areas
• Transport network poor

ZIMBABWE
• Political unrest
• Agricultural land mismanaged

Figure Q3.4.2 Factors affecting development

Study Figure Q3.4.2 above. Explain how both physical and human factors can affect levels of development for some African countries. *(5 marks)*

Intermediate 2 2004 5b

Questions and *Answers* continued ➤

Questions *and* Answers *continued*

Answer

Physical factors can affect levels of development for some African countries because drought for example means people do not have enough water to drink (✓). This causes illness which means they can't work (✓). This means that they and their families don't have enough money to buy food (✓).

Human factors can affect the level of development for some African countries because war for example means people have to leave their homes and belongings to find a new home. This means that they are not bringing money or food home for their families (✓). If they die their family could starve or if homes are attacked people are injured and without medical attention they could die (✓), crops are destroyed (✓). When the fighting is over they have to start all over again (✓).

Comments and marks obtained

This answer correctly discusses both physical and human factors otherwise it would not be possible to gain full marks. Reference to physical factors gains marks for statements on drought, effects of drought, illness and lack of money for food. References to war, causing hardship, starvation, loss of crops and having to begin again are valid points. The answer gains full marks **5 out of 5**. However a more secure answer would have mentioned more physical factors such as relief and climate having an impact on agriculture and transport infrastructure.

Key Point 5

You should be able to identify the main diseases of EMDCs and ELDCs including heart disease, cancer, asthma, AIDS, malaria, cholera and kwashiorkor. You should also be able to describe their distribution and their causes. Figure 3.4.3 shows the world distribution of some of these diseases.

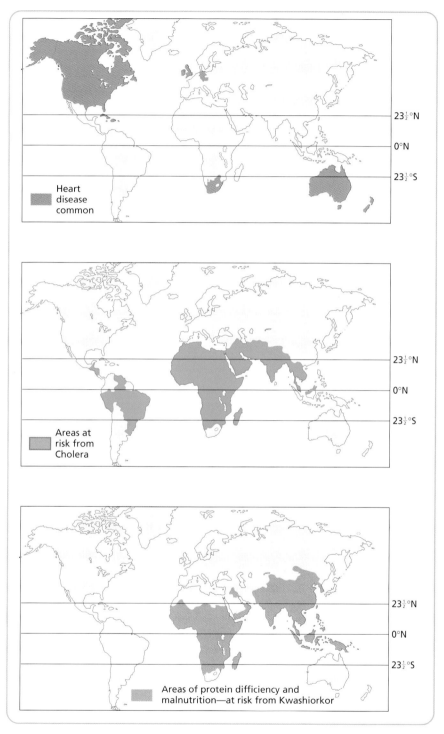

Figure 3.4.3 World distribution of diseases

By studying Figure 3.4.3 you should be able to see certain patterns in the distribution of these diseases which should give some clues as to their causes.

General causes of diseases in EMDCs

Diseases such as heart disease, cancer and asthma are more typical of EMDCs.

◆ Heart disease and cancer have similar causes, much of which is related to environment, lifestyle and genetic factors.

◆ There are many different types of cancer which affect different parts of a body.

◆ Lifestyle factors such as fatty diets, lack of exercise and stress can contribute to heart disease and cancer.

◆ Medical research has proved that smoking is a major cause of lung cancer and heart disease. Smoking can result in lung cancer which can spread to other organs of the body.

◆ Lifestyle such as high consumption of alcohol can cause cancer of the liver and mouth.

◆ Exposure to a badly polluted environment (near heavy industry for example) has been suspected of being a cause of some cancers such as leukaemia (cancer of the blood).

◆ Environmental factors such as air pollution (from sources such as traffic exhausts, industrial smoke), living close to electricity lines, food additives and the use of chemical fertilisers all affect people's chances of contracting cancer, heart disease and asthma.

◆ Family genes are inherited and are a major factor in people suffering from heart related illnesses such as angina, coronary disease and heart failure. Some believe that hereditary factors are important in some cancers.

General causes of diseases in ELDCs

Diseases such as cholera, malaria and Kwashiorkor occur more frequently in ELDCs.

◆ A lack of development, lack of adequate food supplies, widespread poverty, poor standards of hygiene, lack of clean water, poor housing conditions, poor standards of sanitation and local climate conditions all contribute to the occurrence of these diseases.

◆ The population of less developed countries are much more likely to contract and die at an early age from diseases such as cholera, malaria and Kwashiorkor than those diseases associated with the more developed countries of the world.

◆ AIDS is distributed throughout the world. In both more and less developed countries the main cause of the disease is lifestyle. The disease is spread through unprotected sexual contact between infected sufferers and non-affected persons.

◆ Poverty is a major issue in the African continent where AIDS is widespread. Many African countries lack the money and facilities to launch health education programmes or provide the population with preventative methods such as condoms or needle exchange programmes.

◆ Other factors include infected blood supplies, drug users using infected needles and unborn children contracting the disease from infected mothers.

Key Point 6

You should be able to describe the human and physical factors which contribute to the spread of one disease from each of the following lists. (You cannot select AIDS from both lists.)

List A	List B
Heart disease	Malaria
AIDS	AIDS

Some of the general causes have been noted earlier. For heart disease and AIDS most of the factors contributing the spread of the disease are human factors. Heart disease is not an infectious disease whereas AIDS can be passed from one infected person to another.

Factors affecting the spread of heart disease

◆ People whose parents or grandparents suffered from heart disease have a greater chance of developing the disease than those whose family members have not suffered from coronary conditions.

◆ Heavy smoking can lead to the formation of blood clots which can result in heart attacks.

◆ Diets which include high cholesterol content, fatty foods and excess alcohol can lead to blocked arteries (angina) which can cause heart failure in extreme cases.

◆ People who are overweight and fail to exercise may be more prone to heart disease.

◆ Stress from various sources can lead to high blood pressure which can result in heart attacks.

Consequences of the disease for the population in an area you have studied can include:

◆ loss of workdays in industry through ill-health;

◆ increasing costs of state and private healthcare;

◆ need for more medical staff and hospital beds to treat those suffering from the disease;

◆ increase in number of patients suffering from coronary related illnesses such as angina, high blood pressure and strokes;

◆ increasing costs to government to supply subsidised prescriptions of drugs used to control disease symptoms;

◆ lowering of life expectancy rates.

Strategies used to control heart disease

◆ Health education programmes to advise the public on how to avoid heart disease.

◆ Advice on healthy living programmes, the dangers of misuse of alcohol, smoking, poor diet, lack of exercise and dealing with stress.

◆ Prescription of drugs to treat coronary related problems such as high cholesterol, high blood pressure, angina and thrombosis.

◆ Heart clinics to monitor and test for possible heart problems.

◆ Hospital treatment including preventative and curative surgery such as angioplasty, heart by pass procedures and heart surgery.

Effectiveness of strategies

The effectiveness of the above educational and health strategies varies from country to country.

◆ In many parts of Europe people adhere to advice on healthy diets such as in Scandinavia and Netherlands.

◆ In other countries such as Scotland and Germany, fatty diets, heavy drinking and smoking are common and a high percentage of the population suffer from and are prone to heart disease.

Factors affecting the spread of AIDS

◆ AIDS sufferers can infect other people through unprotected sexual contact.

◆ Some people have been infected through receiving blood transfusions of blood which is infected with the HIV virus.

◆ Illegal drug users can infect each other if they share needles with contaminated blood carrying the AIDS virus.

◆ The spread of the disease is more rapid in countries which lack the financial resources to take preventative methods such as education and advertising programmes, issuing contraceptives for protection, setting up needle exchange programmes for drug users.

◆ The disease has reached almost epidemic proportions in parts of Africa whereas in more developed countries measures to contain the disease have been more effective.

Strategies used to control the spread of AIDS

◆ Continuing medical research into finding a cure or vaccine for the AIDS virus.

◆ Use of cocktails of drugs to slow down the progress of the disease of those infected.

◆ Education programmes to educate potential sufferers of the main causes of the disease and preventative measures which can be taken.

◆ Issue of contraceptives for protection.

◆ Setting up of needle exchange centres for drug users to avoid multiple use of infected needles.

◆ Offering financial aid to less developed countries to provide health and education programmes for their populations.

Effectiveness of strategies

◆ In more developed countries such as Britain, USA and western European countries, the rate of infection from AIDS has slowed down due to some of the measures outlined above. Raising awareness has helped people in these countries to take precautions to avoid being infected.

◆ In less developed parts of the world, the rate of infection is rapidly increasing due to poverty, culture, lifestyle and a general lack of medical and educational facilities.

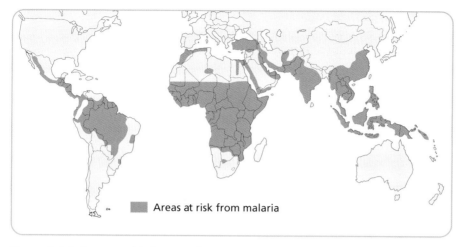

Figure 3.4.4 Areas at risk from malaria

Factors affecting the spread of malaria

◆ Malaria is spread by the female anopheles mosquito. The mosquito is a *vector*, or carrier, of the disease.

◆ They transmit the disease through taking meals of blood from infected persons and pass it on in the next blood meal through their saliva.

◆ These mosquitoes breed in stagnant water such as marshlands under certain climatic conditions, (generally hot, wet climates) with a minimum temperature of 16°C.

◆ The disease can spread very rapidly throughout an area unless certain measures are taken to limit and control this spread.

◆ Mosquitoes have become resistant to many insecticides including DDT and malaria itself has adapted to become resistant to certain drugs which were formerly used to cure it.

◆ As yet there is no vaccine available to prevent infection although great efforts are being made to produce one in medical research facilities.

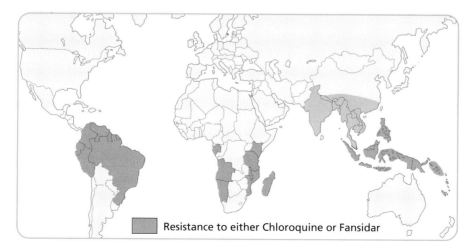

Figure 3.4.5 Resistance to selected anti-malarial drugs

Strategies used to control malaria

◆ Drainage of areas with stagnant water (swamps) and water management schemes to destroy breeding grounds of mosquitoes near rivers.

◆ The use of insecticides such as malathion.

◆ The use of nets to protect people while sleeping from mosquito bites.

◆ The use of drugs to control the disease like quinine or derivatives of this drug (chloroquin). The use of village health centres and issue of information/education through primary health care schemes.

◆ Releasing water from dams to drown larvae.

◆ Using egg white sprayed on to stagnant surfaces to suffocate larvae.

◆ The introduction of small fish in paddy fields to eat larvae.

◆ The planting of eucalyptus trees to absorb moisture.

◆ The application of mustard seeds into water areas which drag larvae below water surface and drown them.

Effectiveness of strategies

The various strategies have met with varying degrees of success.

◆ No effective vaccines have been produced although several test studies in China are achieving progress.

◆ Much depends on local population applying themselves to suggested precautions and taking medication regularly.

◆ Malaria remains an important debilitating and killer disease in many parts of the developing world.

Questions and Answers

Question 3.4.3

'In parts of Africa, village health workers live and work among their people. Their first job is to share knowledge.' Consider the role of the village health worker. Do you think that education is the best way of controlling disease? Give reasons for your answer by referring to a disease that you have studied.   *(4 marks)*

Intermediate 2 2004 5d

Answer

Yes education is a great way of controlling the disease. I think this because if people are educated and are made aware of the disease they can focus on preventing the disease (✓) and keeping themselves safe from potential diseases (✓).

Through education the disease is controlled before it can hurt the victim. People can learn about the disease and exactly what to do to prevent it before it gets a chance to infect people.

Questions and Answers continued ➢

Questions and Answers continued

Answer continued

A disease such as malaria can be controlled through education as people learn that insect repellents, vaccines and by simply closing windows at night (✓) and by keeping themselves covered at night and day they can control the disease (✓).

Comments and marks obtained

This is a good answer which deals well with the different parts of the question. Marks are gained for recognising and explaining the benefits of health education, and for explaining ways in which education has helped control the spread of malaria. The answer obtains **4 marks out of 4.**

Question 3.4.4

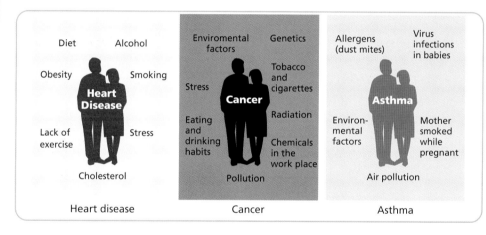

Figure Q3.4.4 Causes of disease

Study Figure Q3.4.4. Select one disease from the following:

heart disease; cancer; asthma. Using the appropriate diagram, explain the ways in which various factors can cause your chosen disease. *(4 marks)*

Intermediate 2 2002 5d

Answer

Heart disease – HEBS (Health Education Board for Scotland) advertise (✓) and show how exercise can beat it (✓). Show how eating more healthily with less fat and alcohol (no beer, wine, spirits) keep you healthy (✓). Foods are also being made more healthy and contain vegetable instead of animal oil and more fruit and vegetables are available and cheap (✓). These have been quite effective but Scotland still has the highest rate of heart disease in Europe (✓).

Questions and Answers continued ➤

HOW TO PASS INTERMEDIATE 2 GEOGRAPHY

Questions *and* Answers *continued*

Comments and marks obtained

The answer makes good use of the data shown in the diagram on heart disease. However, the candidate does not simply lift information. The first statement on HEBS shows good knowledge of the topic and gains the first mark. Further marks are obtained for the references to exercise, diet, food content and the final comment on general effectiveness, noting that Scotland still has the highest death rate in Europe from the disease. There are sufficient points to gain 5 marks, but the question has only 4 marks available. Therefore the answer is worth **full marks.**

Glossary   *Health and Development*

AIDS: Acquired Immune Deficiency Syndrome, which results from sufferers contracting the HIV virus.

Bilateral aid: Aid given from one country to another. It may be financial or given in goods or services such as agricultural equipment or training schemes.

Development: The level of economic and social status of a country.

Development indicators: Measures used to show whether a country is more or less economically developed.

Disease: The process by which a body becomes infected by a foreign body such as a virus or bacteria causing the person to become ill. Some diseases can be passed on by coming into contact with infected persons – contagious – or through contact with a source of infection – food, blood, water. Other diseases such as cancer and heart disease are neither contagious or infectious.

GNP: Gross National Product of a country. It measures the amount of goods and services produced by a country in any given year. Effectively it measures the overall wealth of a country.

Human Development Index (HDI): An indicator which combines social and economic factors which give a fuller definition of a country's development status.

Physical Quality of Life Index (PQLI): Another example of an indicator which combines social and economic factors to describe levels of development.

Section E
Environmental Hazards

The regional context in which this topic is studied is global. Case studies can be taken from any suitable area.

Key Idea 1

You should know what is meant by a natural hazard. Natural hazards are events caused by natural forces which can greatly affect landscapes and people in the areas in which they occur, bringing death, injury and great destruction.

Key Point 1

You should have a detailed knowledge of a tropical storm. You should know and understand their distribution and their general causes.

Distribution of tropical storms

Tropical storms are essentially very deep depressions with windspeeds varying from 60 km/hr to over 200 km/hr. They are located in oceans within 30° of the Equator and begin on the eastern side of the oceans moving westwards before dying out over land. When a tropical storm reach speeds of greater than 120 km/hr they are described as hurricanes. Figure 3.5.1 shows the distribution of tropical storms and Figure 3.5.2 illustrates the main features of this hazard.

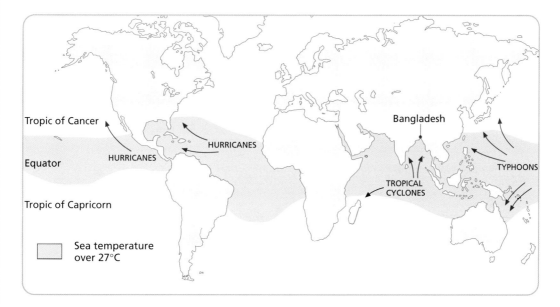

Figure 3.5.1 Distribution of tropical storms

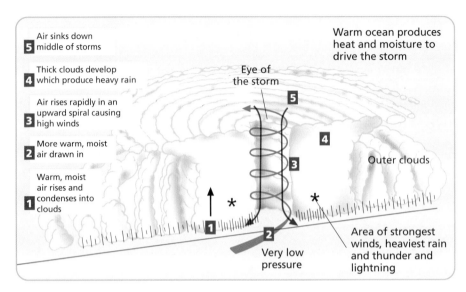

Figure 3.5.2 Features of a tropical storm

Main features and causes

◆ The storm forms over the ocean in the Tropics during the summer. High temperatures cause strong evaporation in deep ocean areas where the hot surface water reach temperatures over 27°C.

◆ Warm moist air rises rapidly, cools and condenses forming very deep cumulonimbus clouds with heavy rainfall.

◆ Low pressure develops and as air is sucked into the depression, high winds develop, increasing from gentle speeds to speeds of at least 60 km/hr.

◆ The rotation of the earth encourages violent winds to rotate around the central 'eye'. These winds are anti-clockwise in the northern hemisphere.

◆ In the central 'eye', the pressure is very low with descending winds, increasing temperature, clear skies and calm, dry conditions.

◆ The whole storm moves forward very quickly, bringing torrential rain, very high winds, falling temperatures and huge clouds in the areas immediately before and after the 'eye'.

◆ Coastal areas over which the storm passes may experience storm surges due to the low pressure causing the sea level to rise with huge waves forming in shallow coastal waters. Low-lying coastal areas are particularly vulnerable to massive damage and loss of life.

◆ As the storm moves away, the pressure rises, winds decrease, temperatures rise and rain turns to showers.

Key Point 2

You should have a detailed knowledge of either an earthquake or a volcano. You should know and understand its distribution and general causes.

Earthquakes

◆ Earthquakes occur when rocks deep within the Earth's crust move suddenly. This movement causes shock waves to travel outwards in different directions through the crust.

◆ The source of the shock waves is called the *focus* and the point immediately above the focus is known as the *epicentre*. This is where the most severe shockwaves occur.

◆ There are three types of shock wave:

● Push or primary waves, 'P' waves, cause rocks to move up and down and are the fastest;

● Shake or secondary waves, 'S' waves cause sideward movement of rocks and move at about two thirds the speed of 'P' waves;

● The third group are known as long waves or 'L' waves which move along the Earth's surface and although they are the slowest, they are much more destructive.

◆ The strength of these waves can be measured on an instrument called a seismometer and are recorded on seismographs. The scale used to describe strength is called the Richter Scale.

◆ The distribution of earthquakes is very similar to that of volcanoes. Both occur along or near the boundaries of the large crustal plates which make up the Earth's crust. These areas are called the plate margins. The Earth's crust is weakest at the margins of the plates.

◆ There are three types of plate boundaries: constructive; destructive; and sliding. At these boundaries the plates are trying to move in different directions due to the fact that the plates are 'floating' on top of the Earth's mantle layer.

Constructive boundaries

◆ In areas of constructive boundaries the plates are forced in opposite directions causing rocks in these areas to be put under a great deal of tension.

◆ Eventually the rocks break and move sharply causing shock waves to travel through to the Earth's surface. These waves cause the ground to shake, creating an earthquake.

Destructive boundaries

◆ In areas of destructive boundaries one of the plates is being forced down beneath another. Great friction is created due to this and the friction stops the plates from moving.

◆ As the pressure continues to increase, the crust eventually suddenly moves downwards into the mantle causing shock waves which create earthquakes on the surface.

Sliding boundaries

◆ In areas of sliding boundaries, crustal plates slide past each other. This sliding movement causes immense friction between the plates.

◆ As the pressure continues to build over time, it overcomes the friction and suddenly one plate moves quickly past the other. This causes shock waves which in turn create earthquakes on the surface.

These movements are summarised in Figure 3.5.3.

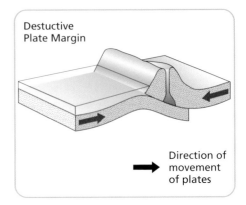

Destuctive Plate Margin

Direction of movement of plates

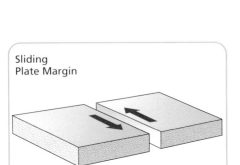

Sliding Plate Margin

Figure 3.5.3 **Earth movements**

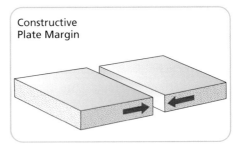

Constructive Plate Margin

Volcanoes

Volcanoes may be described as active, dormant or extinct.

◆ Active volcanoes are those which are likely to erupt.

◆ Dormant volcanoes may not have erupted for some time but could in the future.

◆ Extinct volcanoes are those which will never erupt again.

Distribution

◆ Most volcanoes are found near the boundaries of the crustal plates as shown in Figure 3.5.4.

The most active volcanoes are located throughout the Mediterranean area, the edge of the Pacific Ocean and in the middle of the Atlantic Ocean. Figure 3.5.5 summarises the main causes of volcanoes.

◆ Volcanoes occur where magma, ash, gas and water are allowed to erupt on to the land and sea bed due to a weakness in the Earth's crust.

◆ These weaknesses are most likely to occur at the plate margins, especially at destructive and constructive margins or in areas where the plate is particularly thin, known as 'hot' spots.

◆ Pressure which builds up over long periods of time at the plate margins may be finally released by a volcanic eruption in which liquid magma is forced up through joints and weaknesses of the crust.

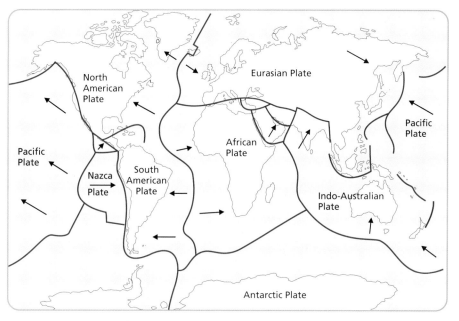

Figure 3.5.4 Crustal plates boundaries

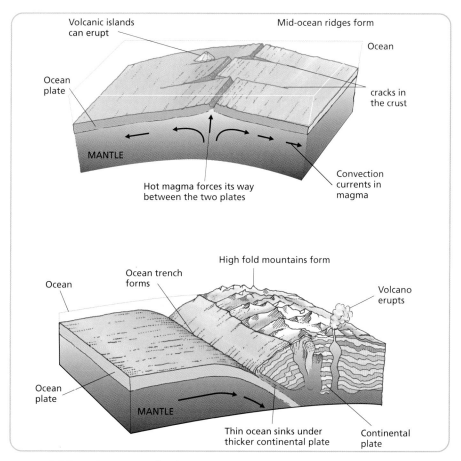

Figure 3.5.5 Main causes of volcanoes

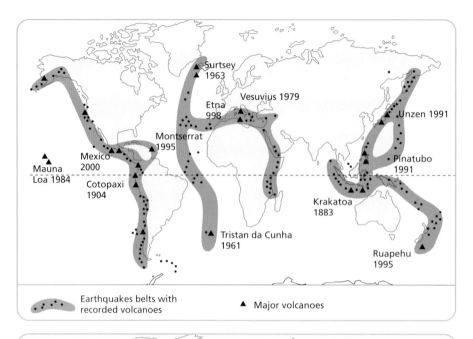

Earthquakes belts with recorded volcanoes   ▲ Major volcanoes

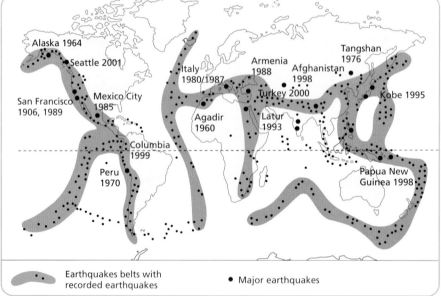

Earthquakes belts with recorded earthquakes   ● Major earthquakes

Figure 3.5.6 Distribution of volcanoes and earthquakes

Questions and Answers

Question 3.5.1

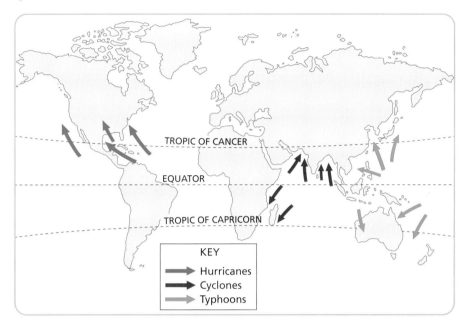

TROPIC OF CANCER

EQUATOR

TROPIC OF CAPRICORN

KEY

➤ Hurricanes
➤ Cyclones
➤ Typhoons

Study Figure Q3.5.1 above. Describe and explain the distribution of tropical storms.
(6 marks)

Intermediate 2 2003 6 b

Answer

Tropical storms are found in close distribution to the tropics of Cancer and Capricorn (✓).
Mainly off the southern coast of North America and the coasts of Australia, South Africa,
China and India (✓). This is due to the high temperatures of the water found in these areas
(over 27 degrees C) (✓) and the depth of this hot water (✓) (at least 6 metres) which
causes the water to evaporate and rise in the atmosphere where it condenses (✓) and
starts to spiral due to the spinning of the Earth. This creates a low pressure so other air is
sucked in over the ocean (✓) to replace it creating strong winds (✓). The process occurs
until the storm reaches land where it loses energy as it has no more warm air to fuel it (✓),
so eventually just disappears.

Comments and marks obtained

Key statements are made on the location of tropical storms and the reasons for their
formation including: high temperatures and depth of water, causing evaporation;
spiralling air which creates low pressure areas; air creating winds and finally loss of
energy. Note that this provides both description and explanation. Had the answer
been based on only one of these it would not have obtained full marks. The answer
has more valid points than is necessary for **full 6 marks.**

Questions and Answers ?

Question 3.5.2

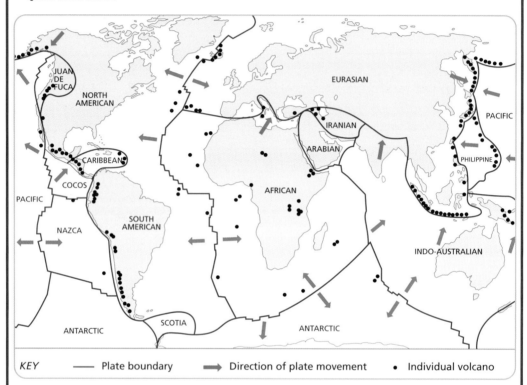

Figure Q3.5.2 Distribution of volcanic activity

Study Figure Q3.5.2. Explain what causes volcanic activity and why it occurs only in certain areas around the world. *(4 marks)*

Intermediate 2 2004

Answer

Volcanoes can occur at either the constructive or destructive plate boundaries. Two crustal plates are slowly moving away from each other (✓) causing molten rock at the centre of the Earth to rise up to the surface (✓) where it cools forming new land. At the destructive plate boundaries one plate is being forced under another as they move towards each other (✓). The rock is exposed to intense heat, pressure and friction as it is pushed towards the centre of the earth and so it melts to form magma (✓). This is forced up through cracks in the Earth's surface resulting in a volcanic eruption (✓). The magma cools on reaching the surface to form solid lava rock again (✓). This process is repeated many times forming a large mountain of layered rock (✓).

Questions and Answers continued ➤

Questions and *Answers* *continued*

Comments and marks obtained

The answer provides a number of valid statements on where and why volcanoes occur. Reference to constructive/destructive plate boundaries, the movement of the plates causing molten rock to rise gains 3 marks. Comments on plate movements, the formation of magma, magma moving through cracks, the cooling of the magma and the layered rock formation are sufficient for a further 4 marks. The answer gains **full 4 marks out of 4.** In fact there are sufficient points to score 7 marks.

Key Point 2

For a tropical storm and either an earthquake or volcano you have studied, you should know the underlying causes and the impact on the landscape and population.

Underlying causes

Depending on which case study you have studied, you should know the detailed causes of the hazard. For example for a particular tropical storm such as Hurricane Mitch which occurred in Central America in 1998, you should know:

◆ how and where the storm formed;

◆ how it developed over a period of time;

◆ the strength of the storm;

◆ the path which it took from beginning to end.

For either an earthquake or a volcano which you have studied in class, such as the Kobe earthquake in Japan in 1995, or the Mount St. Helens volcanic eruption in north west USA in 1980, you should be able to discuss:

◆ the underlying causes of these hazards in some detail;

◆ the type of plate margin at which the hazard occurred;

◆ where the focus and epicentre of the earthquake was;

◆ the strength of the earthquake and;

◆ the extent of the area affected by either the earthquake or the volcano.

Impact on the landscape and population

For tropical storms in your study area you should be able to describe:

◆ the effects of the winds and heavy rainfall on local forests, agricultural areas, soil damage, local buildings and bridges and communications;

◆ the impact of storm surges in coastal areas and their impact on the physical and human landscape

For earthquakes or volcanoes in your study area you should be able to describe:

◆ damage caused to the landscape through earth movements, such as damaged buildings, roads and infrastructure and secondary impact such as fire and flooding, people made homeless, injuries and deaths;

◆ damage caused by lava, ash bombs, ash deposits, mudflows and landslides, the number of deaths caused and the cost of the damage.

Questions and Answers

Question 3.5.3

For a specific tropical storm you have studied:

(i) describe the effects of this natural hazard on the people. *(3 marks)*

(ii) describe what was done to reduce its impact. *(3 marks)*

Intermediate 2 2002

Answer

(i) *A hurricane, which has affected Florida and the Caribbean, usually occurs in the hurricane season which is around September (✔). It is caused by the seas heating up (✔) and hot air rising which swirls around and around. (✔)*

(ii) *The hurricane destroys trees and (✔) and causes huge tidal waves which cover the coasts of Florida (✔). Rain is heavy and mud-slides are common. People sometimes die when their houses collapse from the strong gales (✔). In the Caribbean where they are fairly poor they do not have the money to repair damages caused by the hurricane (✔). Most of their houses are wooded and so would fall down easily in the gales (✔).*

Comments and marks obtained

The first part of the answer correctly identified when storms form, and two causes – seas heating up and hot air rising for **three marks.**

The second part discussed five valid ways in which the hurricane impacted on the landscape and people – destruction of trees, tidal waves, houses collapsing, lack of money for repairs and the houses made out of wood, was more than enough for **full three marks.**

Key Point 3

For each type of natural hazard studied you should know about methods of prediction and their effectiveness.

For **tropical storms**, you should be able to discuss methods such as:

◆ monitoring and tracking storms by satellite and aircraft;

◆ issuing storm warnings in good time;

◆ evacuation plans and procedures;

◆ personal preparation by local people such as building storm shelters, storing food and water supplies and boarding up windows;

◆ preparation of emergency services.

In ELDCs measures may include:

◆ building up and strengthening of banks along rivers and coastlines;

◆ programmes to educate people against risk;

◆ aid programmes from EMDCs which might provide additional protection.

For **earthquakes and volcanoes**, measures may include:

◆ careful monitoring of the situation using appropriate instruments;

◆ watching out for increases in physical activity such increased earthquake activity, gas emissions, changes to the shape of volcanoes which indicate a build-up of pressure;

◆ ground temperature changes measured by satellite;

◆ use of ultrasound to detect magma movements;

◆ planning evacuation and other emergency measures to be put into operation quickly and effectively;

◆ measures to reduce the impact of the earthquake such as strengthening buildings, using shock absorbers in the foundations and structure of buildings, changes in building design, flexible gas, water and power lines;

◆ education programmes;

◆ a good communication system to alert local people of impending disaster in plenty of time.

Key Point 4

You should be able to comment on the effectiveness of the various methods of prediction.

Despite the fact that hurricanes can be predicted in advance, the effectiveness of predictions in minimising the effects is quite limited. Hundreds of people can still be killed. Hurricanes Ivan and Jeanne which affected the Caribbean in August and September 2004 killed several hundred people and caused billions of dollars worth of damage. The predictions are more effective in saving lives and property in more developed areas, where large scale evacuations can be undertaken.

Earthquake predictions are less accurate and effective since earthquakes can happen very suddenly. Many thousands of people have been killed in countries such as Pakistan, Iran, Mexico, China, Russia and Turkey, despite the availability of equipment such as seismometers which can be used to monitor activity.

Predictions about the eruption of volcanoes have been slightly more effective in preventing loss of life. Many people have nevertheless been killed by unexpected volcanic eruptions in various parts of the world.

Predicting the occurrence of hazards are only effective if warnings are heeded and appropriate measures such as evacuation can take place in plenty of time before the disaster occurs.

Key Point 5

You should be able to discuss the role of aid agencies and comment on their effectiveness.

Aid and aid agencies

Aid between countries takes a variety of forms.

◆ Financial aid is given for assistance with building infrastructure, and funding health and education programmes.

◆ Food aid is given in cases of emergency (such as drought and natural disasters) or when a country cannot grow or buy sufficient food for its population.

◆ Equipment, medical supplies and medical assistance can be provided. The medical assistance can consist of doctors, nurses and other medical staff.

◆ Specialist workers can assist in education, training and development programmes in agriculture, industry, education and health care.

Aid is provided as short- or long-term aid, and can be given as voluntary, bilateral, multilateral and tied aid.

Short-term aid

Short-term aid is required immediately following a natural disaster such as a flood or earthquake.

◆ This aid may consist of emergency water supplies, food, medicine, temporary shelter such as tents, rescue operations, emergency power supplies, military labour to help damaged areas, emergency police.

◆ This aid lasts for only a short time but can be crucial in the saving of lives, preventing the spread of disease and giving care and attention to those injured in the disaster, left homeless or without food and water.

◆ This aid does not provide a long-term solution to the problems caused by the disaster.

◆ Much of this aid comes from voluntary organizations such as the Red Cross and Médicin sans Frontières. Often victims are put into refugee camps in which living conditions can be extremely difficult.

Long-term aid

Long-term aid is given to countries over a long period of time to help them develop. This kind of aid is used to help improve the infrastructure of developing countries. The distribution of the aid and the improvements may take a long time to be felt. It does not bring immediate relief following a major disaster but it will bring long term benefits to developing countries.

◆ Aid may be in the form of financial assistance such as loans or grants, personnel assistance with medical and educational programmes, and military aid for peace-keeping activities in war zones.

◆ Major projects such as multi-purpose dams may be built to provide electricity and water supply and irrigation.

◆ Schools and hospitals may be built. Agriculture may be given financial assistance to help farmers improve yields and learn new farming techniques.

◆ Industry may be established in rural and urban areas to improve employment opportunities.

◆ Sewage, sanitation and the provision of clean water may be seen as a priority to improve health standards throughout the country.

Voluntary aid

Voluntary aid is usually given through charities such as Oxfam, the Red Cross, or the Scottish Catholic International Aid Fund.

Multilateral aid

Multilateral aid is aid from international organizations such as the United Nations through development programmes.

Bilateral aid

Bilateral aid is assistance given by one country directly to another country. There are often conditions associated with the way the aid is used.

ENVIRONMENTAL INTERACTIONS

Tied aid

Tied aid is aid given on the condition that the country receiving the aid must spend a proportion of the aid funds to buy equipment or expertise from the donor country. There may also be political conditions attached to the aid.

Problems associated with aid

Although it is designed to help the receiving country, there may be conditions attached to the aid which create problems. Loan repayments can leave the receiving countries repaying far more than they borrowed in the first place. This reduces the ability of the receiving country to use its own money for its own development and reduces the overall effectiveness of the aid.

Debt associated with loan repayments is one of the major problems facing poorer parts of the world and this may continue for many years to come.

Questions and Answers

Question 3.5.4

Select an example of either a volcanic eruption or an earthquake you have studied. Naming the place affected by this event, describe the efforts of charities and relief organisations to reduce its effects. *(4 marks)*

Intermediate 2 2003

Answer

The island of Montserrat was affected by the volcanic eruption of Mount Soufriere in 1997. This island is a British colony and the UK Government offered to provide free transport and £200 (✓) to anyone wanting to relocate to Britain or provide $3840 to people relocating to other islands (✓). A concert was held in London to raise money for the islanders and the UK government offered more money to rebuild the island once the eruption had ended (✓) (£64 Million) (✓). Water purification kits, food and blankets and tents were sent to the island for the islanders (✓).

Comments and marks obtained

By identifying an appropriate study area the answer can be given full marks. The answer identifies four valid ways in which the authorities dealt with the disaster: free transport, £200, relocation funds, and concerts to raise aid and examples of the type of aid sent to the island. The answer therefore gains **4 marks out of 4.**

Questions and Answers

Question 3.5.5

For any tropical storm you have studied describe the long term aid schemes which could be put in place to try to reduce the problems caused by the storm. *(4 marks)*

Intermediate 2

Answer

A storm which I studied is Hurricane Ivan which hit the Caribbean islands in September 2004. This storm caused many deaths and enormous damage. Many of the deaths and much of the damage could have been prevented if long term aid projects had been available. This might have include measures to strengthen the foundations and fabric of houses and other buildings (✓). If money had been spent on educating the population on safety measures such as evacuation (✓) or building shelters (✓) perhaps more lives could have been saved.

Long term aid could also involve improvements to transport and communication e.g strong bridges (✓) to help people evacuate more easily (✓).

Comments and marks obtained

This is a good answer which unfortunately takes some time to get to the important points such as strengthening buildings, educating the population, building shelters and communication improvements which assist evacuations. These points merit 4 marks giving a total of **4 marks out of 4.**

Glossary *Environmental Hazards*

Active volcano: Volcanoes which have erupted recently.

Core: The inner sphere of the Earth.

Craters: Features at the top of a volcano through the build-up of lava deposits.

Crust: The outer layer of the Earth.

Dormant volcanoes: Volcanoes which have not erupted for a long time – several hundreds of years.

Earthquake: A vibration of the Earth's crust caused by shock waves travelling from sudden movements deep within the Earth's crust.

Epicentre: The point on the Earth's surface which is immediately above the source of the shock waves which cause earthquakes.

Extinct volcanoes: Volcanoes which are no longer areas of tectonic activity since records began.

Eye: An area of very low pressure found at the centre of a tropical storm.

Friction: Force of resistance caused by two plates rubbing together and is largely responsible for earthquakes and volcanic eruptions.

Focus: The source point of the shockwaves which cause earthquakes.

Fold mountains: Mountains created by pressure from within the Earth pushing up rocks.

Hurricanes: Another name for a tropical storm.

Lahars: Mudflows formed from a mixture of ash and water travelling at great speeds down the mountain sides during a volcanic eruption.

Lava: Molten rock which pours out of a volcano on to the Earth's surface during an eruption.

Long-term aid: Aid intended to be used over a period of years to help a country develop industry, farming, transport system, education and health care systems.

Magma: Molten rock which lies below the Earth's crustal layer.

Mantle: Layer which lies immediately below the Earth's crust, between the crust and the core.

Multilateral aid: Aid given from a group of countries through agencies such as the United Nations to poorer countries.

Official aid: Bilateral or multilateral aid given to a country.

Pangea: 200 million years ago the Earth's landmasses were joined together to form one single continent known as the Pangea.

Plates: The crust is divided into seven large and twelve small sections known as plates. There are two types of plate – oceanic and continental.

Glossary *Environmental Hazards continued*

Plate margins: Lines on the Earth's surface where two plates meet. There are three types of margin: destructive; constructive and conservative.

Plate tectonics: The theory which explains the movement, formation and destruction of the plates which make up the Earth's crust.

Richter Scale: A measure of the strength or magnitude of an earthquake.

Seismographs: Recordings of the intensity of shock waves within the Earth's crust.

Seismometer: A device used to measure and record shock waves in an earthquake.

Short term aid: Aid given immediately to help an area recover from a major disaster such as a flood, drought, famine, or earthquake.

Stress: The intensity of forces caused when plates move at different speeds in different directions and along with friction is responsible for volcanic activity.

Tectonic activity: Movement which leads to earthquakes, volcanoes and fold mountains and is associated with plate margins.

Tropical storms: Areas of extremely low pressure formed over water areas with temperatures of over 27°C which can travel at speeds of up to 150 km/hr.

United Nations Organisation: A worldwide organization of which nearly all of the world's countries are members. It has a wide range of functions such as World Health, World finance, peacekeeping forces and various aid agencies.

Voluntary aid: Aid given through charitable organizations such as Red Cross or Oxfam or Save the Children Fund.